IV (VI)th All-Russia Scientific-Practical Conference "Prospects of Development and Challenges of Modern Botany"

BIO Web of Conferences Volume 11 (2018)

Novosibirsk, Russia
8 – 12 October 2018

Editors:

E.V. Banaev
A.P. Belanova
Y.G. Zaytseva

ISBN: 978-1-5108-7270-7

Printed from e-media with permission by:

Curran Associates, Inc.
57 Morehouse Lane
Red Hook, NY 12571

Some format issues inherent in the e-media version may also appear in this print version.

Printed by Curran Associates, Inc. (2018)

For additional information, please contact EDP Sciences – Web of Conferences at the address below.

EDP Sciences – Web of Conferences
17, Avenue du Hoggar
Parc d'Activité de Courtabœuf
BP 112
F-91944 Les Ulis Cedex A
France

Phone: +33 (0) 1 69 18 75 75
Fax: +33 (0) 1 69 28 84 91

contact-edps@webofconferences.org

Additional copies of this publication are available from:

Curran Associates, Inc.
57 Morehouse Lane
Red Hook, NY 12571 USA
Phone: 845-758-0400
Fax: 845-758-2633
Email: curran@proceedings.com
Web: www.proceedings.com

TABLE OF CONTENTS

BIO Web of Conferences **11**, 00001 (2018) https://doi.org/10.1051/bioconf/20181100001
Prospects of Development and Challenges of Modern Botany

Effect of a biogenic silica and green-tea-flavonoids-based mechanocomposite on *Fragaria* × *ananassa* Duch. leaf anatomy in *in vitro* conditions

Elena Ambros[1,*], *Victoria* Batrakova[2], *Alexander* Krasnikov[1], *Yulianna* Zaytseva[1], *Elena* Trofimova[3], and *Tatyana* Novikova[1]

[1]Central Siberian Botanical Garden, Siberian Branch of the Russian Academy of Sciences, Zolotodolinskaya st., 101, 630090 Novosibirsk, Russian Federation
[2]Novosibirsk State Pedagogical University, Vilyuyskaya st., 18, 630126, Novosibirsk, Russian Federation
[3]Institute of Solid State Chemistry and Mechanochemistry Siberian Branch of the Russian Academy of Sciences, Kutateladze st., 18, 630128 Novosibirsk, Russian Federation

Abstract. For the first time light and scanning electron microscopy have been to reveal changes in the anatomical structure of leaf blades of *Fragaria* × *ananassa* Duch. regenerants cultivated in nutrient media supplemented with biogenic silica and green-tea-flavonoids-based mechanocomposite. An increase in the density of stomata per unit area of the leaf, size of the stomatal gaps, thickness of the epidermis and mesophyll have been noted. The revealed changes in the leaf blade structure under mechanocomposite treatment contribute to the further successful acclimatization of regenerants to *ex vitro* conditions.

1 Introduction

Strawberry (*Fragaria* × *ananassa* Duch.) is the most widely consumed berries throughout the world. *F. ananassa* is traditionally propagated vegetatively by rooted runners. However this method is not suitable for production of strawberry cultivars due to accumulation and transfer of viral, fungal and bacterial diseases from mother plants. The use of clonal micropropagation allows to obtain not only a healthy planting material, but to increase significantly the rate of commercial reproduction of *F. ananassa* cultivars. Despite the effectiveness of the technology, *in vitro* cultivation causes a number of anomalies that manifest at the level of functional morphophysiological features of the regenerant leaf blades, such as underdevelopment of the epidermis and photosynthetic tissues, including the palisade parenchyma, vascular bundles, and formation of an abnormal stomatal apparatus with a small number of stomata per 1 mm^2 of the leaf surface [1]. All these changes are due to *in vitro* culture conditions characterized by high relative humidity, low gas exchange, presence of exogenous carbohydrates and low rate of CO_2 absorption [2]. As

* Corresponding author: ambros_ev@mail.ru

a result, when transferring microclones to *ex vitro* conditions, decrease of survival rate leading to reduce of technology efficiency is possible. A number of studies have shown that *in vitro* application of silicon compounds improves photosynthetic capacity of plants, positively affects the anatomical and morphological parameters of the leaves [3, 4]. Silicon-containing preparations obtained from renewable plant raw materials by mechanochemical methods arouse considerable interest. The advantage of mechanochemical technology of raw materials processing is extraction of valuable compounds with the least impact on the environment, without additional costs of thermal energy and chemical reagents [5]. Mechanocomposites of this type have increased solubility, retain their natural composition and therefore are most effectively absorbed by plants, which determines the prospects of their use in biotechnologies as new biologically active compounds with growth regulating activity and at the same time, safe for human health and environment [6, 7]. We suppose that application of the mechnocomposite studied may provide opportunities for improvement of physiological adaptation to the transferring stress during ex vitro acclimatization of plantlets. Therefore, the objectives of the present study were (i) to analyze the anatomical features of regenerant leaves under the influence of a mechanocomposite based on biogenic silicium dioxide and flavonoids of plant origin during the rooting stage in *in vitro* conditions and (ii) to optimize the protocol of clonal micropropagation for strawberry large-scale production using the mechanocomposite.

2 Materials and methods

2.1 Characterization of mechanocomposite

A mechanocomposite (MC) was produced from rice husks (*Oryza sativa*, cv 'Liman,' Krasnodar Krai, Russia) and green tea (*Camellia sinensis* L.) (Krasnodar Krai, Russia) at the Institute of Solid State Chemistry and Mechanochemistry of the Siberian Branch of the Russian Academy of Sciences (Novosibirsk, Russia) by mechanical activation in an RM-20 roller mill with water cooling. The rotation speed was 1000 rpm. The material residence time in the active zone was 40–60 s [8]. Rice husk contains high levels of biogenic silica. The leaves of *C. sinensis* contain phenolic compounds with antioxidant properties. These compounds have chelating functional groups and can increase significantly the solubility of silica.

2.2 Plant material and culture conditions

The anatomical structure of micropropagated plantlet leaf blades of two strawberry cultivars, 'Alpha' and 'Solnechnaya polyanka', under the influence of mechanocomposite during the rooting stage in *in vitro* conditions was studied. Microshoots (1.0-1.5 cm) with 2-3 leaves were inoculated on Gamborg-Eveleg's basal salt medium - B_5 [9] supplemented with 20 g L^{-1} sucrose, 6 g L^{-1} Bactoagar (PanReac®, Spain) and different concentrations of MC (2.5, 5.0 and 10.0 mg L^{-1}) for one passage of 8 wk. MC-free B_5 was used as control. The pH was adjusted to 5.5 before autoclaving (121°C, 2.1 atm for 20 min), and mechanocopmosite was added to the medium before autoclaving. The cultures were maintained in culture jars (15 mL medium per vessel) at 23 ± 2°C with a 16-h photoperiod with 40 μmol m^{-2} s^{-1} of light intensity provided by cool white fluorescent lamps (Philips, Pila, Poland).

2.3 Anatomical and stomatographic studies of leaf structure

The leaf anatomy, stomata number, and size were measured in the fresh plant material after 8 weeks. The second leaf from the apical bud of the shoot was taken for analysis. Sections of 50-60 μm were made using an MC-2 sledge microtome with a TOC-II thermo-cooling table (Tochmedpribor, Kharkiv, Ukraine). The histological analysis was performed using a microscope equipped with an Axioskop-40, an AxioCam MRc5 camera, and AxioVision 4.8 software (Carl Zeiss). Morphometric measurements were conducted in 10 fields of vision for each preparation. Stomatal density was calculated on the abaxial side of the leaf using a Hitachi TM-1000 scanning electron microscope with original software (Hitachi High-Tech., Tokyo, Japan). This allowed biological objects to be investigated without spraying. Sections of the middle part of the leaf (5–7 mm^2) from ten randomly selected plantlets of each study group were taken for the analysis. Ten leaves per processing were measured, using one leaf per plantlet. Stomata density was calculated on ten distinct, nonrepeating lines of sight for each leaf. Anatomical features of stomata were determined for 300 randomly selected stomata.

2.4 Statistical analysis

Experiments were repeated three times, and 20 microshoots were used for each treatment. To assess the effect of mechanocomposite on the anatomical parameters, data were subjected to analysis of variance (ANOVA) using STATISTICA™ 8 (StatSoft, Inc., Tulsa, OK). Significance between means was tested by Duncan's new multiple range test (P = 0.05). Data are presented as a means ± SE.

3 Results and Discussion

Estimation of stomatal density in *in vitro* culture showed that this parameter depended on MC concentration in medium and increased significantly up to 60 % in comparison with control (Table 1). According to literature an increase of stomatal density provides a higher absorption of CO_2 [10] positively affecting the intensity of photosynthesis [11]. Maximum of stomatal density was observed on medium supplemented with 5 mg L^{-1} MC (Fig. a, b). Moreover, statistically significant differences in the morphometric parameters of the stomata were revealed. The ratio of the stomata polar diameter to the equatorial diameter characterizes the degree of their closure.

Table 1. Stomatal density and stomata size of leaf abaxial surfaces of *F. ananassa* plantlets under MC after 8 wk of culture

Cultivar	MC, mg L^{-1}	Stomata number per mm^2	Polar diameter (PD), μm	Equatorial diameter (ED), μm	PD/ED
'Solnechnaya polyanka'	0.0	291.75±16.8^{d}	20.48±0.20^{a}	10.91±0.13^{b}	1.93±0.02^{a}
	2.5	312.54±19.3^{cd}	15.22±0.09^{c}	8.88±0.06^{c}	1.74±0.02^{b}
	5.0	469.14±25.4^{a}	17.94±0.17^{b}	11.20±0.13^{b}	1.65±0.02^{c}
	10.0	390.84±23.1^{b}	20.44±0.14^{a}	13.03±0.11^{a}	1.60±0.02^{c}
'Alpha'	0.0	281.13±27.4^{b}	21.22±0.14^{a}	10.56±0.11^{b}	2.07±0.02^{a}
	2.5	336.58±17.1^{ab}	20.61±0.15^{b}	11.31±0.10^{a}	1.86±0.02^{b}
	5.0	374.92±38.1^{ab}	19.93±0.15^{c}	11.27±0.11^{a}	1.83±0.02^{b}
	10.0	364.52±38.5^{a}	15.13±0.14^{d}	8.20±0.14^{c}	1.86±0.02^{b}

Means followed by the same letter are not significantly different (P=0.05)

In vitro stomatal closure results in a low level of gas exchange and oxidative stress [12], which leads to chlorophyll destruction and emergence of vitrified shoots due to low conductivity of gas and water [13].The largest ratio of the polar stomata diameter to the equatorial one was noted in the control plants (1.93 and 2.07, depending on the genotype), the smallest in the variants of media with MC (from 1.65 to 1.86, depending on the genotype). Thus, the addition of MC in the base medium resulted in the formation of stomata with a smaller diameter and a high stomatal density. Such changes in the leaf structural and functional organization under MC contribute to an increase in the adaptability of regenerants to conditions with low humidity and intensive lighting during the plantlet acclimatization to *ex vitro* conditions [14].

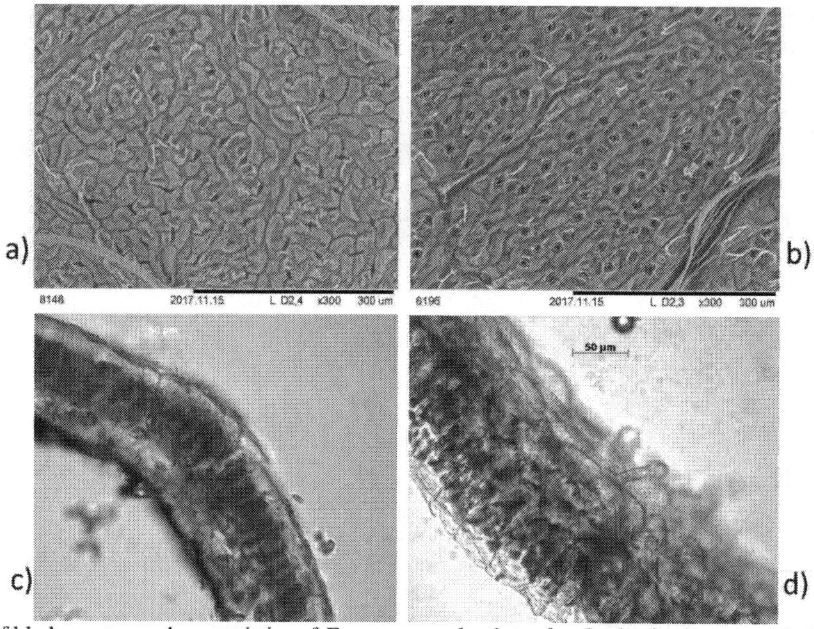

Fig. Leaf blade anatomy characteristics of *F. ananassa* plantlets after 8 wk of culture ('Alpha'): a) stomata on hormone-free B_5 medium; b) stomata on medium supplemented with 5 mg L^{-1} MC; c) transverse section of leaf blade on hormone-free B_5 medium; d) transverse section of leaf blade on medium supplemented with 10 mg L^{-1} MC.

Comparative analysis of the anatomical structure of leaf blades of regenerants cultivated on media with MC revealed that the thickness of the adaxial epidermis increased reliably by 20-30% ($P < 0.05$) under MC treatment in comparison with the control (Table 2). The thickness of the abaxial epidermis of 'Alfa' cultivar was also higher than in control by 30-40% when MC was added to the base medium at different concentrations (2.5-10.0 mg L^{-1}). Statistically significant differences in abaxial epidermis thickness of 'Solnechnaya polyanka' were obtained on media supplemented with only 5.0 mg L^{-1} MC (up to 10%) ($P < 0.05$). In the presence of all the tested concentrations of MC (2.5-10.0 mg L^{-1}), significant differences in the total mesophyll thickness (up to 30%) were noted in 'Alpha' cultivar (Fig. c, d). The established differences were due to the high thickness levels of both the spongy and palisade parenchymas. Statistically significant differences in abaxial epidermis thickness as well as mesophyll thickness of 'Solnechnaya polyanka' were obtained on media supplemented with only 5.0 mg L^{-1} MC (up to 10%) ($P <0.05$).

Table 2. Effect of MC on the thickness of leaf tissues of *F. ananassa* plantlets after 8 wk of culture

Cultivar	MC, mg L^{-1}	Thickness, μm				
		adaxial epidermis	abaxial epidermis	mesophyll	palisade parenchyma	spongy parenchyma
'Solnechnaya polyanka'	0.0	18.2±0.24^{c}	21.6±0.21^{b}	103.5±0.67^{b}	49.9±0.24^{c}	53.6±0.65^{a}
	2.5	20.1±0.29^{b}	21.9±0.16^{b}	107.9±0.89^{ab}	53.1±0.48^{b}	54.9±0.51^{a}
	5.0	20.9±0.20^{b}	23.3±0.15^{a}	113.4±1.30^{a}	56.7±0.48^{a}	56.8±0.99^{a}
	10.0	22.1±0.21^{a}	22.3±0.25^{ab}	107.2±2.43^{b}	52.5±1.02^{b}	54.7±1.85^{a}
'Alpha'	0.0	15.8±0.31^{b}	16.3±0.44^{c}	71.6±1.83^{c}	36.2±1.12^{c}	35.4±0.98^{d}
	2.5	20.6±0.62^{a}	23.0±0.63^{a}	90.2±1.67^{a}	47.2±1.19^{a}	42.9±0.63^{bc}
	5.0	20.7±0.47^{a}	23.3±0.62^{a}	79.8±2.15^{b}	41.1±1.25^{b}	38.7±1.70^{cd}
	10.0	20.8±0.61^{a}	21.2±0.82^{b}	91.6±4.86^{a}	43.9±2.02^{ab}	47.6±3.04^{a}

Means followed by the same letter are not significantly different (P=0.05).

Chlorophyll as it is well known accumulates mainly in the mesophyll cells of the leaf. The quantitative changes observed in the increase of mesophyll thickness under MC treatment can be regarded as a feature providing increase of the leaf photosynthetic activity of plants. Thickening the epidermal tissues demonstrates the positive response of regenerants to MC application and can stimulate the plant resistance to adverse environmental conditions providing mechanical protection.

4 Conclusions

Thus, the thickness of the chlorophyll-bearing parenchyma and epidermal tissues of regenerant leaf blades of two *F. ananassa* cultivars is increased under MC treatment, which indicates the formation of plants characterized by a higher photosynthetic potential. The changes noted contribute to the further successful acclimatization of regenerants to *ex vitro* conditions. The results obtained on the application of MC based on biogenic silica and flavonoids of plant origin can be used for development of commercial systems for the production of *F. ananassa* improved planting material using biotechnological approaches.

In vitro propagation of *F. ananassa* microplants was carried out with the financial support of the budgetary project of the Central Siberian Botanical Garden, SB RAS No AAAA-A17-117012610051-5 within the framework of the State Assignment. Anatomical analyses of plantlets were supported by the Russian Foundation for Basic Research and the Government of Novosibirsk Region as research project No 17-44-540339. In our study, *in vitro* material from the collection of the Central Siberian Botanical Garden SB RAS – USU 440534 "Collection of living plants indoors and outdoors" was used.

References

1. E.F. George, M.A. Hall, G.-J. De Klerk, *Plant propagation by tissue culture* (3rd Edition, Springer, 2008)
2. L. Argita, A. Gonzalez , R.S. Taméz, Physiologia Plantarum **115,** 166-173 (2002).
3. F.T. Braga, C.F. Nunes, A.C. Favero, M. Pasqual, J.G. Carvalho, E.M. Castro, Pesqui Agropecu Bras **44**, 128-132 (2009)
4. X. Yao, J. Chu, K. Cai, L. Liu, J. Shi, W. Geng, Biol. Trace Elem. Res. **43,** 507-517 (2011)
5. O.I. Lomovsky, I.O. Lomovskiy, D.V. Orlov, Green Chemistry Letters and Reviews **4,** 171-185 (2017)

6. O.A. Rozhanskaya, O.I. Lomovsky, E.G. Trofimova, T.G. Lomova, Success mod. Sci. Educ. **8**, 31-37 (2016)
7. E.V. Ambros, S.Y. Toluzakova, L.S. Shrainer, E.G. Trofimova, T.I. Novikova, In Vitro Cell. Dev. Biol. – Plant **54**, 436-443 (2018)
8. E.G. Shapolova, Lomovsky O.I., Patent of Russian Federation, No. RU 2438344 C1, Byull. Izobret. **1** (2012) (in Russian)
9. O.L. Gamborg, D.E. Eveleigh, Can. J. Biochem. **46**, 417-421 (1968)
10. E.M. Castro, F.J. Pereira, R. Paiva, *Histologia vegetal: estrutura e função de órgãos vegetativos* (UFLA, Lavras, 2009)
11. Y.M. Zhou, S.J. Han, Photosynthetica **43**, 445-449 (2005)
12. K. Asada, Annu. Rev. Plant Physiol. Plant Mol. Biol. **50**, 601-639 (1999)
13. A. Krieger-Liszkay Singlet, J. Exp. Bot. **56**, 337-346 (2005)
14. S.A. Asmar, E.M. Castro, M. Pasqual, F.J. Pereira, J.D.R. Soares, Sci. Hortic. **161**, 328-332 (2013)

Climatic modeling of the distribution of *Oxytropis triphylla* (Fabaceae) by maximum entropy method

Konstantin Baikov[1,*], and *Denis* Krivenko[2]

[1]Central Siberian Botanical Garden SB RAS, Novosibirsk, Russia
[2]Siberian Institute of Plant Physiology and Biochemistry SB RAS, Irkutsk, Russia

Abstract. Predictive spatial models of the distribution of *Oxytropis triphylla* (Pall.) Pers. (Fabaceae), an endemic species of Baikal Siberia, were generated in MAXENT computer program using maximum entropy method. Long-term data of air temperatures for every month of the year were downloaded from the world database of open access WorldClim. Modeling was performed separately for minimum, average and maximum temperatures. Each variable contribution to the modeling was the basis to select the key variables having higher influence on the obtained models. The selected 10 key variables are the following: minimum temperatures of December and January; average temperatures of October, December, January and February; maximum temperatures of November, December, January and February. Then a model of the second level was calculated using only the ten key variables. There are three northern localities in the zone of adverse temperature effects: cape Malyi Cheremshanyi, Chencha and Sakhuli villages (all of them are in the Republic of Buryatia). It has been experimentally confirmed that the values of the key variables along the coasts of the Maloe More of Lake Baikal (Irkutskaya Oblast) are the most favorable for habitation of *O. triphylla* in this part of its range.

1 Introduction

The results of modeling the distribution of plants and animals according to climatic and bioclimatic variables by the maximum entropy method have been published in many foreign journals as well as in Russian scientific publications in the last few years [1–3]. Interest to the implementation of such studies is due to the high information content of the predictive models obtained. In this study, the maximum entropy method was used to conduct a quantitative analysis of the climatic temperature factors that limit the range of *Oxytropis triphylla* (Pall.) Pers., the endemic species of Baikal Siberia [4, 5].

2 Materials and Methods

* Corresponding author: kbaikov2018@mail.ru

Specimens of *O. triphylla*, stored in the herbarium collections of Central Siberian Botanical Garden of SB RAS (Novosibirsk, Russia), Siberian Institute of Plant Physiology and Biochemistry of SB RAS (Irkutsk, Russia) and V.L. Komarov Botanical Instutute of RAS (Saint-Petersbourg, Russia) were taken as the materials for the study. For the construction of the predictive models, 45 localities were chosen. Accurate geolocation references were determined for these localities. The predictive spatial models were generated in the MAXENT computer program by the maximum entropy method [6–9]. Values of climatic indicators were downloaded from the world database of open access WorldClim – http://worldclim.org/.

3 Results and Discussion

Initially, climate models were constructed separately for minimum, average and maximum temperatures of each month of the year. The variables contribution to the modeling was the basis to select the set of the key variables having higher influence on the models obtained. The selected ten key variables are the folowing: minimum temperatures of December and January; average temperatures of October, December, January and February; maximum temperatures of November, December, January and February. Then a model of the second level was calculated using only the ten key variables (Fig. 1).

Fig. 1. The predictive spatial model of *Oxytropis triphylla* range constructed on 10 key temperature indicators. Explanations are given in the text.

In Fig. 1, the areas with the temperature regimes which determine the probability of finding *O. triphylla* within the interval of 0.93–1.00, or 93–100% are coloured in red; with probability 0.86–0.92 are coloured dark orange; with probability 0.78–0.85 – orange; and 0.70–0.77 – yellow. Thus, red-orange-yellow gamma indicates the favorable parts of Baikal

Siberia where this species may be found. Different tones of the colour blue indicate areas that are unfavorable for habitats of the species: in the dark blue areas the probability of finding varies in the range 0.00–0.08; in the pure blue areas – 0.09–0.15; in the light blue – 0.16–0.23; in the very light blue – 0.24–0.31.

Different tones of the green colour indicate areas with the probability of finding the species from 0.32 to 0.69. Green coloured areas may be considered as neutral and even close to favorable for the species. White squares correspond to geographical points where the herbarium specimens were collected and included in the modeling (45 localities). Some of them are located very closely to each to other; thus, they look as the same white square.

According to the obtained predictive spatial model, the main fragment of the range of *O. triphylla* is located in the middle part of the Western coast of Lake Baikal (38 localities). Therefore, it can be considered experimentally confirmed that values of the key temperature variables near the coasts of the Maloe More of the Lake Baikal (Irkutskaya Oblast) are the most favorable for habitation of *O. triphylla* in this fragment of its range. Accordingly, new populations of this endemic species may be found in this territory.

Another range fragment includes only three localities in Southern Buryatiya. These localities are placed a long distance from the main part of the range described above (150–200 km). Regimes of key temperature indicators in Southern Buryatiya allow predicting a high probability to find *O. triphylla* in a new neighbouring locations.

Three northern localities (cape Malyi Cheremshanyi on the Western Coast of Lake Baikal, Chencha and Sakhuli villages – all in the Republic of Buryatia) are disposed far from the main part of the range. According to the predictive spatial model in Fig. 1, these populations inhabit the zone of adverse temperature effects, especially in the neighborhood of Chencha and Sakhuli villages.

The contribution of variables to the construction of the predictive spatial model for *O. triphylla* is presented on the column diagram in Fig. 2, where the dark blue colour columns show the value of the variable when it is taken into account independently of the others; the light-blue colour columns show the result of excluding the variable from the prediction across the entire set of variables; the lower (red) column demonstrates the total contribution of all the variables to the constructed model.

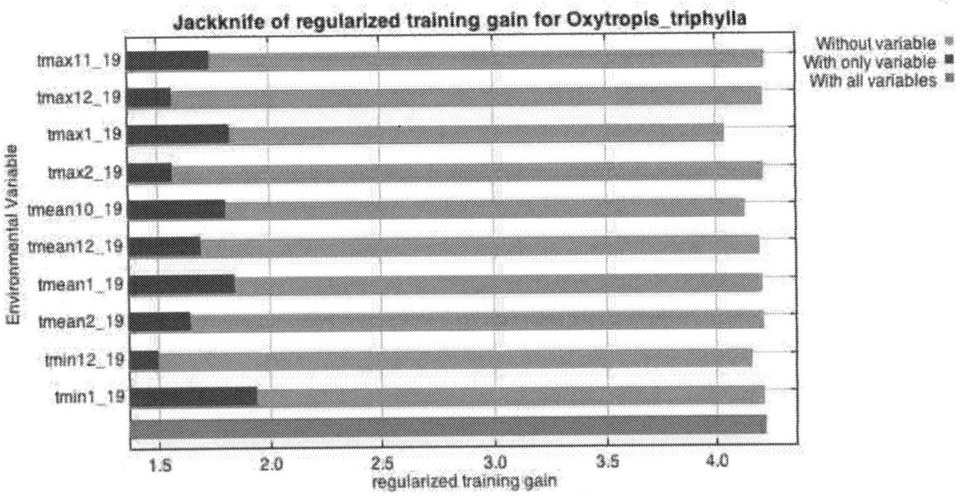

Fig. 2. The diagram of contribution of key temperature indicators to the spatial predictive model of *Oxytropis triphylla*.

According to the individual contribution, the three leading variables are the following (in descending order of values): the minimum temperature in January (tmin1_19), the average temperature in January (tmean1_19) and the maximum temperature in January (tmax1_19).

Consequently, favorable temperature regimes in January largely determine the parts of the range with a high probability of *O. triphylla* finding. The average temperature in October (tmean10_19) and the maximum temperature in November (tmax11_19) also have high significance. The elimination of any variable from the full set of indicators leads to decreasing the training role of the full set, the most noticeable after excluding the maximum temperature of January (tmax1_19), the minimum temperature of December (tmin12_19) and the average temperature of October (tmean10_19).

The work was carried out within the framework of the state task of the Central Siberian Botanical Garden of SB RAS for the project AAAA-A17-117012610055-3 "Biological diversity of cryptogamous organisms (algae, fungi, lichens) and vascular plants in the geospace of biotic and abiotic factors, assessment of their role in water and terrestrial ecosystems of Northern Asia" with the partial support of the Russian Foundation for Basic Research (project No. 16-04-01399A). In preparing the publication, the materials of the bioresource scientific collection of CSBG SB RAS "Herbarium of Higher Vascular Plants, Lichens and Fungi (NS, NSK)" UNU No. USU 440537 were used. The research was done using the collection of the Core Facilities Center "Bioresource Center" at the Siberian Institute of Plant Physiology and Biochemistry SB RAS (Irkutsk, Russia).

References

1. D.V. Sandanov, Uchenye zapiski Zabaikalskogo Gosudarstvennogo Universiteta **11**, 38–45 (2016)
2. D.V. Sandanov, B.B. Naidanov, *Rastitelnyi Mir Aziatskoi Rossii* **3**, 30–35 (2015)
3. S. V. Solodyankina, E. A. Istomina, A. A. Sorokovoi, V. V. Chepinoga, Geography and Natural Resources **5**, 92–99 (2016)
4. L. I. Malyshev, *Genus Oxytropis* (Conspect of the flora of Asian Russia: Vascular plants, Novosibirsk, 2012)
5. D.G. Chimitov, O.V. Imetkhenova, B.B. Naidanov, *Rastitelnyi Mir Aziatskoi Rossii* **1**, 10–18 (2017)
6. S.J. Phillips, R.P. Anderson, R.E. Schapire, Ecol. Modeling **190**, 231–259 (2006)
7. S.J. Phillips, M. Dudik, Ecography **31**, 161–175 (2008)
8. S.J. Phillips, M. Dudik, R.E. Schapire, *Proceedings of the 21st Intern. Conf. on Machine Learning* (ACM, New York, 2004)
9. J. Elith, S. J. Phillips, T. Hastie, et al., Diversity Distrib **17**, 43–57 (2011)

Influence of salinity (NaCl) on the photosynthetic pigments content of some sweet sorghum varieties

Gulnaz Baiseitova[1,*], *Batyrbek* Sarsenbayev[2], *Erlan* Kirshibayev[2], and *Madiar* Kamunur[1]

[1]Kazakh National Agrarian University, Almaty, 050010, Republic of Kazakhstan
[2]RSE «Institute of Plant Biology and Biotechnology», Almaty, 050040, Republic of Kazakhstan

Abstract. Identification and introduction of crops into the production that are sufficiently adapted to drought and salinization of soils capable of providing high productivity with limited water consumption is becoming increasingly important currently, and in particular in prospect under the circumstances of aggravating of climate aridization and desertification processes in the Central Asian region. These requirements are met by sorghum, the main economic features of which are universality of use, exceptional heat,- drought,- and salt tolerance, as well as high productivity. The article presents the results of studying the effect of a saline nutrient medium on the content of photosynthetic pigments in some varieties of sweet sorghum to identify the most resistant varieties to chloride salinity. It was shown that the concentration of a and b chlorophylls, as well as carotinoids in leaves decreases with increasing NaCl concentration. The content of photosynthetic pigments under conditions of salinity was higher in resistant varieties than in unstable varieties. The quantitative content of photosynthetic pigments affects the intensity and productivity of photosynthesis which determines the yield of agricultural crops.

1 Introduction

Saline soils occupy 20% of the world's developed lands, and half of all irrigated lands are highly salinized. The tendency of soil salinity growth persists. For example, 900 million hectares in the world are valued as salinized land which accounts for approximately 6% of all world soils or about 20% of the world's developed territories [1, 2].

The irrational use of natural resources leads to environmental disruption. The total area of Kazakhstan is 272 million hectares, and of these 180 million hectares is currently threatened with degradation which is 60% of the total territory of the country. Degradation is accompanied by intensive salinization of the soil which leads to an increase in the territory of saline desert areas in the internal drainage basins and salinization of irrigated lands [3]. The number of cultivated plants that can grow on saline soils is insignificant. One such plant is sorghum.

* Corresponding author: b.g.naz@mail.ru

Sorghum is a unique cereal plant, both for its biological characteristics and for its economically valuable characteristics. Its main advantages are exceptional drought tolerance, salt tolerance, high productivity, stable yields by year, good fodder dignity and universality of use [4, 5, 6, 7].

Sorghum tolerates soil salinity better than other crops [8,9]. This culture is able to grow normally and develop at a salt concentration in the soil two times higher than that required by corn. This culture, perhaps, can be cultivated on almost all soils except wet and cold soils with a close standing of groundwater [10].

Sorghum is a good meliorating crop when sown on alkali soils and a reliable remedy for combating secondary salinization. Thus, variety crops not only provide high yields of grain and green mass, but also take out salts from the soil from 31 to 75 tons per hectare, including harmful ones, such as chlorides and sulfates. Sorghum forms a large biomass of plant residues which ultimately enriches the soil with organic substance and serves as a phytomeliorant [11].

The cultivation of sweet sorghum in the Republic of Kazakhstan, to a considerable extent, may save irrigation water (resource-saving), restore soil fertility (desalinization, improve water-physical properties), improve ecology (regulation of atmospheric humidity and temperature, O2/CO2 balance), provide the population with liquid sugar, livestock farms with forages, solve the problems of energy in the region (biofuel).

In connection with the abovementioned, we conducted a comparative assessment of sugar sorghum varieties for salt tolerance in order to identify and select the most promising varieties for cultivation on saline soils in southeast Kazakhstan. The quantitative content of photosynthetic pigments was used as criteria for salt tolerance.

2 Materials and Methods

Domestic and foreign varieties (15 varieties) of sweet sorghum (Sorghum saccharatum Pers.) were taken as the object of research. The plants were grown 10 days in solutions containing various salt concentrations: control, 0.3%, 0.6%, 0.9% NaCl. Determination of photosynthetic pigments was carried out using the Shlyk method [13]. Extraction of chlorophylls «a» and «b», as well as carotinoids from leaves was carried out by the alcohol extraction method, and their concentration was found by spectrophotometric method (SF-Genesys 10 uv) at 649 (644) and 665 (662) nm, carotinoids at 440,5 nm. The pigment content was calculated by the formulas (Werner formulas):

$$CA\ (mg/l) = 11{,}63 \times D665 - 2{,}39 \times D649$$

$$CB\ (mg/l) = 20{,}11 \times D649 - 5{,}18 \times D665$$

The content of carotinoids was determined by the Wettstein's formula:

$$Ccar\ (mg/l) = 4{,}695 \times D440{,}5 - 0{,}268 \times (CA + Bmg/l)$$

Conversion of the pigment content to mg/g was carried out according to the formula:

$$A = C \times V / P \times 1000$$

where A is the pigment content, mg/g; C is the concentration of chlorophyll, mg/l; V is the volume of drawing, ml; P is the weight, g.

All research results were subjected to statistical processing by Udolskaya [14] and computer programs Excel 97, ANOVASYSTAT 2007.

3 Results and discussion

A study of the content of photosynthetic pigments in sweet sorghum varieties showed that a weak concentration of NaCl (0.3%) stimulated the formation of chlorophyll a in the leaves of varieties Kazakhstanskoye-16, Sever, Stavropol-36, Larets, Borotala and Sazhen. High salt concentrations (0.6-0.9%) suppressed the formation of chlorophyll a more in varieties Rostovskyi, Oranzhevoye-160, Sakharnoye-32 (Figure 1).

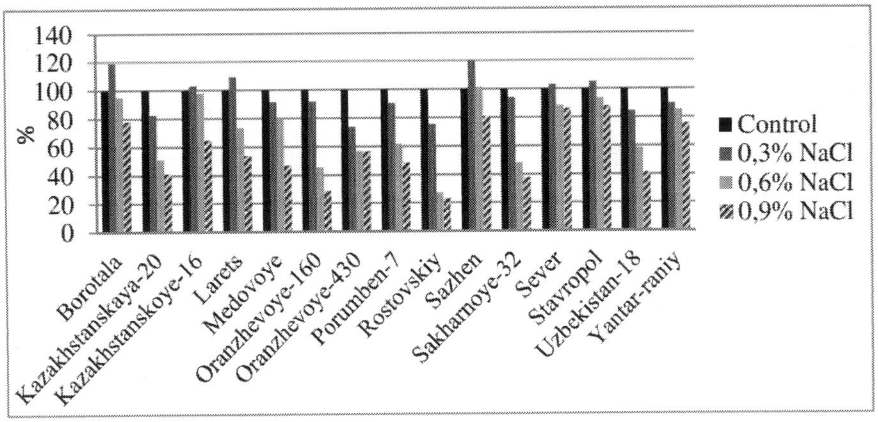

Fig. 1. Influence of salinity on the content of chlorophyll a in leaves of sweet sorghum varieties.

Similar results were obtained when analyzing the content of chlorophyll *b* in leaves of sweet sorghum varieties. Weak salinity caused an increase in the content of chlorophyll *b* in varieties Sazhen, Stavropol-36, Uzbekiston-18 and Borotala. A sharp decrease in the pigment content under the influence of salt was found in the Rostovskyi variety. The content of this pigment was less affected by salinity in the varieties Sever, Stavropol-36 and Uzbekistan-18 (Fig. 2).

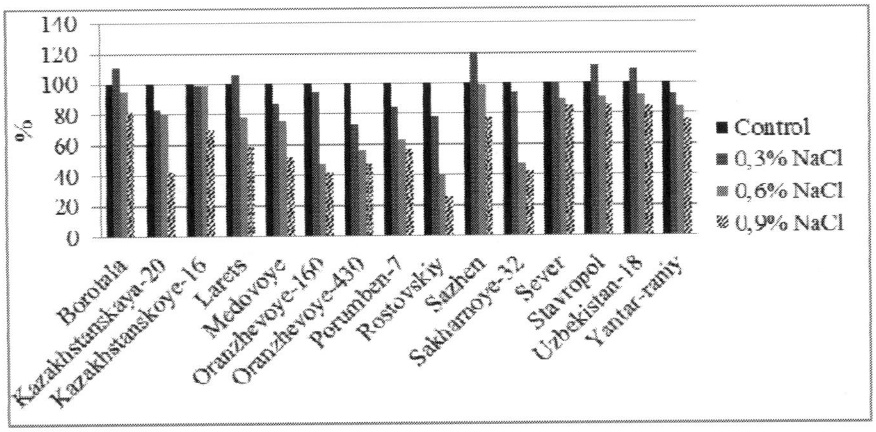

Fig. 2. Influence of salinity on the content of chlorophyll b in leaves of sweet sorghum varieties.

Carotinoids are involved in various defense mechanisms: due to the presence of conjugated double bonds. They are able to bind singlet oxygen and inhibit the formation of free radicals preventing their negative effect on the organism. They can also act as antioxidants protecting sensitive tissues and labile compounds from oxidation. It can be assumed that adverse environmental factors can induce the synthesis of carotinoids, especially in stress-resistant samples.

Indeed, a weak concentration of chloride salinity increased the content of carotinoids to 17-18% in varieties (Figure 3).

Sazhen and Stavropol-36 differ in relative resistance to salinity. The content of this pigment increased to 5-6% in varieties Larets and Sever which also were resistant to salinity.

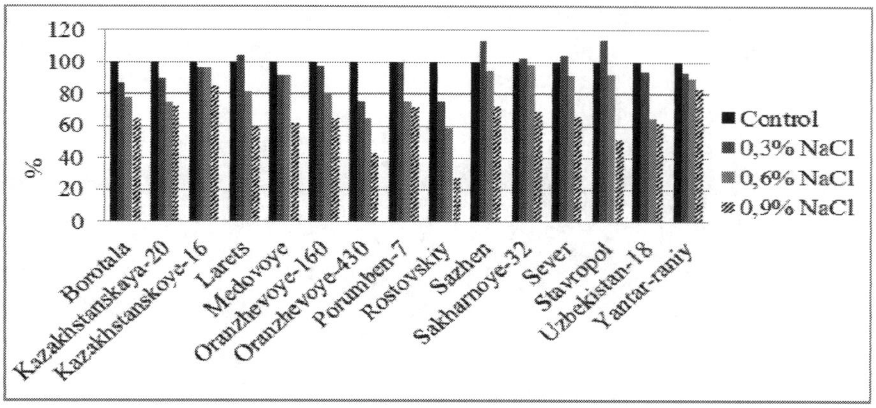

Fig. 3. Influence of salinity on the content of carotinoids in leaves of sweet sorghum varieties.

Thus, this characteristic, the quantitative content of carotinoids can be used as a marker of resistance of varieties to salinity.

As a result of studies of 15 different varieties of sweet sorghum we detected relatively salt-resistant forms for the content of photosynthetic pigments that can be grown in slightly saline soils, and can also be used as a starting material for obtaining more stable forms of sweet sorghum.

References

1. T. Flowers, Journal of Experimental Botany **55**, 307-319 (2004)
2. R. Munns, M. Tester, Ann. Rev. Plant Biol. **59**, 651-681 (2008)
3. M.A. Orlova, A.S. Saparov, *Global self-regulating circulation of salts in nature.* (Almaty, 2009)
4. N.I. Nesmeyanova, *Features of the soil cover and the fertility condition of arable soils in the Samara Region* (Kinel, 2002)
5. A.V. Alabushev. *Adaptive technology of grain sorghum cultivation to the arid zone of the Northern Caucasus* (Zernograd, 2000)
6. A.P. Tsarev E.V. Morozov, *Agrobiological bases of cultivation and use of sorghum crops in the Volga region* (Sarayov, 2011)
7. O.A. Pergayev, Corn and sorghum **1**, 29-32 (2013)
8. Y.I. Isakov, Agriculture of Russia **4**, 14-22 (1977)
9. B.N. Malinovskyi, Steppe areas **9**, 24-28 (1981)
10. L.P. Beltyukov, A.A. Gritsenko. *Application of fertilizers for grain crops in the Don* (Collection of scientific work. Institute of sorghum,Zernograd, 1993)
11. G. Levakhin, Y. Sidorov, N. Dokina, Milk and meat cattle breeding **8**, 20-23 (2010)
12. A.A. Shlyk, *Biochemical methods in plant physiology* (Nauka, Moscow, 1971)
13. N.L. Udolskaya. *Introduction to Biometrics*. (Nauka, Alma-Ata, 1976)

BIO Web of Conferences **11**, 00004 (2018) https://doi.org/10.1051/bioconf/20181100004
Prospects of Development and Challenges of Modern Botany

Seed and biological productivity of *Prunella vulgaris* L. in situ and ex situ

Irina Barsukova [1,*], and *Alexei* Astashenkov[2]

[1]N.F. Katanov, Khakasiya State University, 655017, Abakan, Russia
[2]Central Siberian Botanical Garden SB RAS, 630090, Novosibirsk, Zolotodolinskaya str., 101, Russia

Abstract. Reproductive biology and biological productivity of *Prunella vulgaris* L. were studied in natural growth conditions in Khakasiya and under introduction. It was established that in the wild a potential seed productivity of the species varied from 85,4 to 160,3 it./ a shoot, a real one – from 73,6 to 149,2 it.; erem germination under optimal conditions reached 90,5%. Under introduction the indicators of seed productivity increase by 2-5 times, of erem germination – by 1,5 times. Biological productivity of individuals in cultivation is by 5-10 times higher compared to that in natural habitats.

1 Introduction

Common selfheal (*Prunella vulgaris* L.) is a perennial herbaceous polycarpic plant with sympodially accreting rhizome. The species under study has a circumboreal range [1]. It is a valuable medicinal plant which posesses antimicrobic and antiviral activities [2]. In nature it is low-yielding. In this regard study of reproductive characteristics and biological productivity of *P. vulgaris* in the wild and under introduction will allow not only to judge` adaptation of plants to the environment, but to obtain high-yielding crops. There is a considerable amount of data on reproductive biology of the species, but they encompass Europe and America [3-6]. In Siberian natural conditions the indicators of seed productivity of this species were studied only in Novosibirsk [7] and Tomsk [8] oblasts information on biological productivity is absent. The aim of the work was study and comparison of the pattern of reproductive biology and biological productivity of *P. vulgaris* in natural conditions of Khakasiya and under introduction.

2 Material and methods

Study of the indicators of reproductive biology of *P. vulgaris* was conducted in 2012-2014 in four coenopopulations located in different phytocoenoses of Khakasiya. Coenopopulation (CP) 1 is in bluegrass-orchard grass (*Dactylis glomerata*, *Poa sibirica*) forest meadow ; total projective cover degree (TPCD) 98%, projective cover degree of the species (PCD) 5%. CP 2 is in common lady's mantle (*Alchemilla vulgaris*) true meadow;

* Corresponding author: saphronovairina@mail.ru

TPCD 95%, PCD 10%. CP 3 is in grasses-horsetail (*Equisetum sylvaticum, Poa pratensis*) forest meadow, TPCD – 65%, PCD – 2%. CP 4 is in orchard grass-aizoon stonecrop (*Sedum aizoon, Dactylis glomerata*) true meadow, TPCD – 45%, PCD – 2%. The introduction experiment was performed in Krasnoyrsk Krai. Under cultivation the pattern of reproductive biology was studied on individuals grown from erems collected in the same coenopopulations. Study of seed productivity was conducted by I. V. Vainagiy method [9]. Calculations were carried out on randomly selected ripe generative individuals. Germination pattern was also studied on the erems collected in the wild and under introduction. The erems of typical of *P. vulgaris* natural conditions of habitation (CP 2) were sown under winter to study biological productivity.

3 Results and discussion

The fruit of *P. vulgaris* is a coenobium, it consists of 4 fruitlets – erems of the combined type of dormancy [10]. *P. vulgaris* erems are fine, the mass of 1000 items in coenopopulations does not exceed 0,67±0,02 g. Study of seed productivity indices in coenopopulations allowed to establish their similarity and dissimilarity depending on the year of study and phytocoenosis type (Table 1).

Table 1. Seed productivity of *Prunella vulgaris* individuals in natural habitats (per 1 generative shoot)

№ CP	Year	PSP, it.	V*,%	RSP, it.	V,%	Cp,** %
1	2012	160,32±9,59*** 84-240	29,91	149,24±8,77 76-229	29,38	93,1
	2013	135,68±8,04 72-216	29,65	122,40±8,12 48-207	33,18	90,2
	2014	125,76±6,54 72-192	26,00	111,20±5,97 62-186	26,86	88,4
2	2012	121,28±7,96 72-192	32,82	110,76±7,22 61-181	32,57	91,5
	2013	100,80±6,04 48-168	29,96	85,84±6,46 29-142	37,62	85,2
	2014	111,36±4,97 48-168	22,33	101,08±4,65 40-152	22,98	90,7
3	2012	134,88±8,59 36-216	31,87	124,72±8,17 28-204	32,77	92,4
	2013	120,96±7,65 48-192	31,63	109,44±6,82 36-174	31,19	90,4
	2014	111,52±11,95 24-168	53,58	96,84±10,76 16-156	55,58	86,8
4	2012	112,64±8,03 36-168	35,67	103,76±7,61 24-156	36,69	92,1
	2013	85,44±6,17 36-144	36,11	73,64±4,98 27-111	33,83	86,2
	2014	98,80±6,82 24-144	34,47	87,24±5,50 24-124	31,54	88,2

Note: *Coefficient of variation; **coefficient of productivity (Cp); ***above the line – average value±mistake of average value; below the line – minimal value – maximal value.

The highest values of potential seed productivity (PSP) were in all coenopopulations in 2012, low ones – in 2013 and 2014. The greatest values therewith were noted in 2012 in forest meadows, the lowest ones – in 2013 in true meadows. The distinctions established in PSP indices in individuals were linked with meteorological conditions of the vegetative season. Thus, prolonged moderate humidity and favourable warm weather from May to September in 2012 reflected positively on formation of the reproductive sphere compared to other years. Values of the real seed productivity (RSP) of *P. vulgaris* individuals regardless of habitat conditions were somewhat lower than PSP ones (Table 1), which was linked with not only the ecotope, but with high effectiveness of self-pollination directly affecting erem setting.

When introducing, increase in productivity by 2-5 times takes place independent of ecological-phytocoenotic species growth (Table 2). This is due to essential increase of the number of flowers in the main inflorescence (up to 78) on the shoots and in paracladia (up to 48). In addition, distinctions between reproductive indicators are determined by ontogenetic plant stage. Thus, the greatest values of PSP (up to 688 it.) and RSP (up to 76 it.) are observed in individuals at the ripe generative stage.

Of the relation between seed productivity of *P. vulgaris* and the number of generative shoots and inflorescences on them, as well as the number of flowers in the latter wrote A. A. Winn and P. A. Werner [5]. According to recent research conducted in Finland, the cutting regime may also positively affect reproductive characteristics of the species [11].

Table 2. Seed productivity of *Prunella vulgaris* individuals under introduction (per 1 generative shoot)

№ CP	Year	PSP, it.	V,%	RSP, it.	V,%	Cp, %
1	2012	366,72±11,29 264-480	15,39	352,92±10,90 240-457	15,45	96,2
	2013	499,64±18,23 336-648	18,24	493,52±17,91 331-631	18,15	98,7
2	2012	308,48±9,93 216-384	16,11	297,88±9,47 198-359	15,90	96,6
	2013	547,84±13,01 444-688	11,87	537,24±13,05 436-676	12,14	98,0
	2014	318,68±12,61 144-408	19,78	299,56±12,56 136-388	20,96	94,0
3	2012	306,40±9,88 192-384	16,13	295,44±10,02 187-374	16,97	96,4
	2013	463,84±16,78 336-624	18,09	449,88±16,73 331-615	18,59	97,0
	2014	296,60±15,87 120-408	26,75	283,80±16,12 116-401	28,41	95,7
4	2013	456,20±17,76 336-600	19,47	441,76±17,27 325-598	19,54	96,8

The coefficient of seed productivity in nature varies between 85,2 and 93,1%, under cultivation – between 94,0 and 96,6%. *P. vulgaris* individuals studied by Yu. Kurochkina (2015) in Novosibirsk Oblast are characterized by somewhat lesser values of productivity (Cp. does not exceed 78,7%). A significant coefficient of productivity (90%) is recorded for southern Tomsk Oblast (90%) [8]. High reproductive indicators of *P. vulgaris* cause anxiety at foreign scientists. In evergreen tropical forests of South America the species is evident as invasive, it easily spreads and brings the threat to local flora [12].

Germinating ability and germination energy of erems 6 months after gathering are higher compared to germinating ability of erems kept for 2 years. Percentage of germination is usually higher in erems collected in true meadows. The optimal temperature for germination turns out to be 28-30°C. At such temperature regime germinating ability of seeds after 6 months of storage varies between 62,5 and 90,5%, 2 years after gathering – between 30,0% and 69,5%. As the temperature regime decreases, germination indices decline reaching the minimum at 10-13°C. Germinating ability of seeds collected in culture is somewhat higher (up to 97,5%) than in nature. Germination energy is higher (96,5% on the 10th day of germination).

Biological productivity of *P. vulgaris* individuals in culture is higher in comparison with that in their natural habitats. Individuals at ripe generative ontogenetic stage are characterized by the greatest above-ground mass at sowing density of 100 seeds per 1 running meter. The above-ground mass of raw material obtained under introduction during the second year of vegetation (228,05±36,05 g/run.m) exceeds dry mass of raw material collected in the wild by 5-10 times (g_2–3,17 g/m^2).

4 Conclusion

P. vulgaris is a species of good potential ability of propagation by seeds. In various ecological-phytocoenotic conditions the potential seed productivity of individuals varies from 85,4 it. to 160,3 it., the real one – from 73,6 it. to 149,2 it.; germination ability of erems under optimal conditions of germination varies between 62,5 and 90,5%. Emergence of plantlets in the wild has irregular character and depends on weather conditions of the current year. Under introduction the indices of seed productivity increase by 2-5 times, germination of erems – not more than by 1,5 times.

Biological productivity of *P. vulgaris* individuals in culture is by 5-10 times higher compared to that in their natural habitats (g_2-3,17 g/m^2). Individuals of the second year of development at the ripe generative ontogenetic stage are characterized by the greatest above-ground mass (228,05±36,05 g/run.m)

The study was conducted within the framework of the Project of State Assignment № AAAA-A17-117012610053-9 (CSBG SB RAS, Russia).

References

1. A.G. Borisova, *Flora SSSR* **20** (Academiae Scientiarum, Moscow, 1954)
2. *Vegetable resources of Russia. Wild flowering plants, their component composition and biological activity. Families Caprifoliaceae – Lobeliaceae* **4** (KMK, St. Petersburg, 2011)
3. T.E. Miller, A.A. Winn, D.W. Schemske, Amer.J.Bot. **81**, 1-6 (1994)
4. A.A. Winn, Ecology **73**, 831-840 (1985)
5. A.A. Winn, P.A. Werner, Ecology **68**, 1224-1233 (1987)
6. A.A. Winn, Ecology **69**, 1537-1544 (1988)
7. N.Yu. Kurochkina, Bul. of ASAU **11**, 84-87 (2015)
8. T.G. Harina, N. V. Shvydkaya, *Tomsk oblast nature complex* **2** (TGU, Tomsk, 1995)
9. I.V. Vaynagiy, Bot. J. **59**, 826-831 (1974)
10. M.G. Nikolaeva, M.V. Rasumova, V.N. Gladkova, *Reference book on dormant seed germination* (Nauka, Leningrad, 1985)
11. J. Jantunen, K. Saarinen, A. Valtonen, S. Saarnio, Applied Vegetation Science **10**, 285-292 (2007)
12. O. Godoy, A. Saldaña, N. Fuentes, F. Valladares, E. Gianoli, Biol.Inv. **13**, 1615-1625 (2011).

BIO Web of Conferences **11**, 00005 (2018) https://doi.org/10.1051/bioconf/20181100005
Prospects of Development and Challenges of Modern Botany

Naturalization of *Prunus pensylvanica* L.f. (*Rosaceae*) in Novosibirsk

Anastasiya Belanova[1,*] , and *Ludmila* Chindyaeva[1]

[1]Central Siberian Botanical Garden, Siberian Branch RAS, 630090 Novosibirsk, Russia

Abstract. Naturalization of the North American species *Prunus pensylvanica* has been revealed in Novosibirsk for the first time. This species was introduced in the city in the middle of the last century. It naturally regenerates vegetatively in the area of landscape objects and in dedrological collections and gives self-seeding. In local conditions it is characterized by fast growth, short pregenerative period, presence of abundant uneven-aged progeny, high vegetative mobility, and local population-forming ability.

1 Introduction

Naturalization and distribution of alien plant species are one of the serious ecological problems in many regions of the world [1,2]. At present an active process of distribution of alien species is observed in Siberia [3]. Among naturalizing woody plants in Siberian cities there are representatives of North American flora, including *Prunus* L. species [4]. In Novosibirsk we noted an intensive natural regeneration of *Prunus pensylvanica* L.f. which was introduced in local conditions in the middle of the last century. *P. pensylvanica* regenerates vegetatively in large quantities at places of cultivation, penetrates in open forest stands, gives self-seeding, actively spreads in the area. Closely related North American *Prunus serotina* Ehrh. and *Prunus virginiana* L. have got an invasive status in European countries [5-7].

The aim of our study was assessment of the degree of P. *pensylvanica* naturalization in the second range in Novosibirsk.

2 Materials and methods

Study was conducted in green stands of Novosibirsk and dedrological collections in different districts of the city in 2009–2017. Seed and vegetative generations as individuals of seed origin, root suckers, and partial plants were noted. Trial fields for assessment of natural regeneration of *P. pensylvanica* were laid out. Age status of individuals was determined by T. A. Rabotnov (1987) [8]. The degree of naturalization was estimated with the use of appropriate classification [9].

* Corresponding author: boronina.a@inbox.ru

3 Results and discussion

It was revealed that the alien species *P. pensylvanica* actively regenerated in the area of landscape objects and dendrological collections in Novosibirsk. It forms abundant uneven-aged generation with predominance of fruit-bearing individuals, gradually expands the occupied territory and goes beyond habitats. High vegetative mobility and distribution of daughter plants at a considerable distance from maternal individuals, as well as self-seeding were recorded in *P. pensylvanica*. The species naturally forms large clump-clones on the illuminated sites, in places goes under the canopy of open stands, gives self-seeding on the forest margins. It was established that for the period of introduction in Novosibirsk P. pensylvanica sucessfully adapted in local conditions; it is winter-hardy and characterized by fast growth, stable in the urban environment growing in various ecological conditions – in parks, in the area of residentail development, near streets and roads.

In the arboretum of Siberian Research Institute of Plant growing and Breeding (SRIPGB) *P. pensylvanica*, planted in the amount of 23 specimens in 1989 г. (it was received by 3-year old seedlings from Abakan and seeds from Central Siberian Botanical Garden (CSBG SB RAS), formed a local self-reproducing population. On the plot where 7 maternal trees 3,7 - 8,6 m high survived, natural regeneration of *Prunus pensylvanica* amounted to 63 specimens (1 - 5 specimens per 1 m^2). Flowering and fruit bearing are at the age of 4-5, ground seed germination in local conditions does not exceed 7%. Among daughter individuals prevail generative: young (g1), middle-aged (g2) and old (g3) plants (Fig. 1). Regular single flowering and fruit bearing are noted in young generative individuals up to the age of 12, abundant ones – in middle-aged (20 years old) individuals.

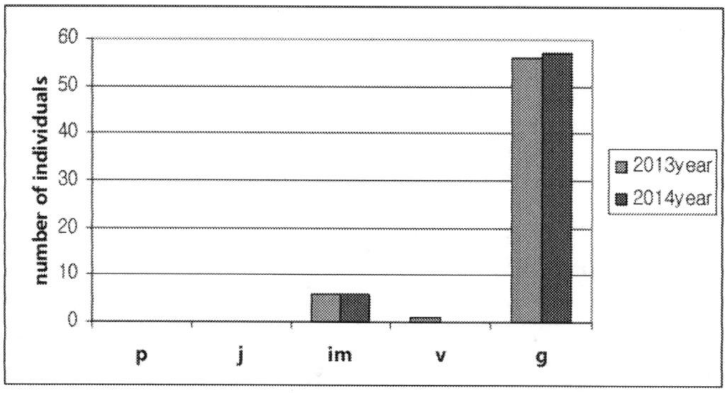

Fig. 1. The age spectrum of *Prunus pensylvanica* local population in the arboretum of SRIPGB: p – seedlings, j – juvenil age stage, im – immature, v – virginal, g – generative stage.

P. pensylvanica in Novosibirsk is not very long-lived, noticeable decrease in annual growth begins in middle-aged generative plants (Fig. 2). Old generative individuals represent the smalest group of fruit-producing plants (less than 5% of population abundance), presence of up to 15–30% of drying out branches and xerocephalous features are characteristic of them. At the age of more than 30, dying off of the natural trunk is observed. Replacement of trunks by fast growing individuals of vegetative origin takes place in short time. In the community with domination of *P. pensylvanica* species diversity of herbaceous cover acounts for 26 species, the projective cover being from 47 to 96 %. In its second range in Novosibirsk the North American species *P. pensylvanica* is characterized by the same biological features which are peculiar to the plants in the natural range. In the natural habitats (western and eastern parts of North America, forest regions of Canada) *P. pensylvanica* is a pioneer species of felling sites, burnt areas and different disturbed territories [10,11].

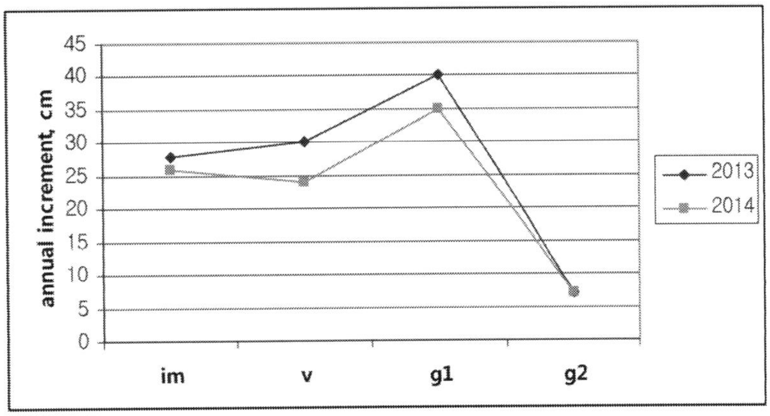

Fig. 2. Dynamics of annual increment *in Prunus pensylvanica* individuals at different age stages: im – immature, v – virginal, g1 – young generative, g2 –middle-aged generative.

The tree age in the wild does not exceed 40 years, fruiting begins at the age of 2, the plants have a shallow root system and give abundant root shoots spreading over a considerable distance. Seeds are charaterized by poor germination (less than10 %), but high seedling viability – up to 100 %, seed germination increases after fires and at succession disturbances [12,13]. In some regions of North America *P. pensylvanica* grows together with closely related *P. virginiana* and *P. serotine* which are characterized by high invasive activity on European continent.

In Novosibirsk *P. pensylvanica* exhibits dynamics of growth and development and reproducive ability typical of the species in natural habitats; it naturally regenerates both vegetatively and by seeds. For the first time self-seeding of *P. pensylvanica* was marked in the city in 1960[14]. In the urban environment the species is ecologically plastic, grows in various conditions, including areas with anthropogenic impact, regenerates more intensively on well illuminated sites. Daughter individuals of various ages occur at places of growing *of P. pensylvanica* maternal plants. The availability of uneven-aged generation of the introduced species testifies to the active process of its naturalization in local conditions [15]. Natural regeneration of *P. pensylvanica* was recorded in different Siberian cities [16, 17].

Such indicators as speed of growth, short pregenerative period, annual abundant fruitage, intensive vegetative regeneration and self-seeding, natural distribution on the territory at places of cultivation, ability to form self-reproducing populations allowed to state for the first time the beginning of naturalization of *P. pensylvanica* (status 3) in Novosibirsk. The data obtained testify to potential risk of the ingress of the alien species in urban forests and different natural communities. In this regard monitoring of the regeneration dynamics and prediction of *P. pensylvanica* spreading over the territory of Novosibirsk to prevent biological invasion of the species in the region is an important task.

4 Conclusions

The original data on natural vegetative regeneration and that by seed of the alien North American species *P. pensylvanica* introduced in Novosibirsk in the middle of the last century were obtained. It was revealed that the species actively regenerated naturally on urban objects and in dendrological collections spreading at a considerable distance from maternal individuals and going beyond places of cultivation. Fast growth, short pregenerative period, availability of abundant uneven-aged generation, high vegetative mobility, ability of colonization of new territories and formation of self-reproducing populations allowed to state for the first time the beginning of naturalization of *P. pensylvanica* in Novosibirsk.

References

1. D.M. Richardson, P. Pysek, PPG. **30**, 409 (2006)
2. Y.K. Vinogradova, S.R. Mayorov, L.V. Khorun, The Black Book of the Flora of Central Russia (Moscow, 2010) in Russian
3. The Black Flora Book of Siberia (Novosibirsk, 2016) in Russian
4. L.N. Chindyaeva, A.P. Belanova, T.I. Kiseleva, RJBI. **2,** 90 (2018)
5. V. Wagner, M. Chytrý, B. Jiménez-Alfaro, J. Pergl, S. Hennekens, I. Biurrun, I. Knollová, C. Berg, K. Vassilev, J.S. Rodwell, Ž. Škvorc, U. Jandt, J. Ewald, F. Jansen, I. Tsiripidis, Z. Botta-Dukát, L. Casella, F. Attorre, V. Rašomavičius, R. Ćušterevska, J.H.J. Schaminée, J. Brunet, J. Lenoir, J.-C. Svenning, Z. Kącki, M. Petrášová-Šibíková, U. Šilc, I. García-Mijangos, J.A. Campos, F. Fernández-González, T. Wohlgemuth, V. Onyshchenko, P. Pyšek, Div. Distr. **23**. 1 (2017)
6. A. Halarewicz, L. Zolnierz, For. Ecol. MaN. **313**, 91 (2014)
7. J. M. Heberling, T. Kichey, G. Decocq, FEC. **6**, 875 (2016)
8. T.A. Rabotnov, *Experimental Phytocenology* (MSU, Moscow, 1987) in Russian
9. A.A. Notov, Yu.K. Vinogradova, S.R. Mayorov, RJBI. **4**, 54 (2010)
10. G.W.Wendel, *Silvics Manual* (published online, 2004)
11. A.F. Rhoads, T.A. Block, *Trees of Pennsylvania: a complete reference guide* (Penn Press, Philadelphia, 2005)
12. P.L. Marks, PEMO. **44**, 73 (1974)
13. L.R. Auchmoody, Can. J. For. Res. **9.** 514–516 (1979)
14. A.V. Skvortsova, Z.G. Ekaterincheva, *Recommendations on the use of introductions in forestry and gardening* (Novosibirsk, 1981) in Russian
15. D.M. Richardson, P. Pysek, M. Rejmanek, M. Rejmanek, M.G. Barbour, F. D. Panetta C. J. West, Div. Distr. **6**, 93 (2000)
16. Z.I. Luchnik, *Decorative durability of shrubs* (Kolos, Moscow, 1988) in Russian
17. A.I. Lobanov, M.I. Sedaeva, N.A. Konovalova, Fruit Grow, Seed Prod, Introd Wood plants. **20**, 96 (2017)

Seed productivity and dynamics of germination of *Ziziphora* L. species in the Pamir-Alay

Kobil Bobokalonov[1,*], and *Vera* Cheryomushkina[2]

[1]Institute of Botany, Physiology and Plant Genetics, AS RT, Dushanbe, Tajikistan
[2]Central Siberian Botanical Garden SB RAS, 630090 Novosibirsk, Russia

Abstract. Seed productivity of 4 *Ziziphora* species: *Z. pamiroalaica, Z. interrupta, Z. brevicalyx,* and *Z. suffruticosa* growing in the Pamir-Alay Mountains was studied. It was revealed that the indicator of real seed productivity did not exceed 40 %. Seed viability and character of germination were studied. It was established that seeds of all species did not have a dormancy period, had high viability, up to 95 % and high germination energy.

1 Introduction

Seed productivity is one of the important indicators of species vitality in different living conditions [1]. It governs a species role in the phytocoenosis and expresses plant adaptation to environment conditions [2, 3]. Study of seed productivity is widely considered in modern literature [4-6].

Ziziphora representatives dominate in the plant cover of the Pamir-Alay and form in the high mountains the florocoenotype "thyme communities" characteristic of Central Asia, they grow on stony-cobble slopes and screes from the low to the high mountains [7, 8]. The species are widely used as medicinal plants [9] and are overharvested in the wild, which leads to population depletion. Study of seed productivity and seed germination will allow to reveal mechanisms of adaptation and sustainable development of species in various communities of the montane belt Pamir-Alay. In this regard the aim of the paper was study of seed productivity and seed germination of 4 *Ziziphora* species growing in the Pamir-Alay Mountains.

2 Material and methods

The material on seed productivity of 4 *Ziziphora* species was collected in 2016-2017 in the territory of Mogoltavsky and Gissaro-Darvazsky floristic regions of the Pamir-Alay (Tajikistan): 1. *Z. pamiroalaica* Juz. – Gissarsky Range, gorge of the Varzob river, vicinities of Ziddi vilage, 2800 m above sea level, tall herbage semisavanna, pebble-clayey

* Corresponding author: kobil_5@bk.ru

slope; 2. *Z. interrupta* Juz. – Karateginsky Range, gorge of the Ramit river, vicinities of Zairon, 1200 m above sea level, shibliak, stony slope; 3. *Z. brevicalyx* Juz. – Khozrati-Shakh Range, vicinities of Lyangar village, 2200 m above sea level, thermophilous juniper stand, pebble-clayey slope ; 4. *Z. suffruticosa* Pazij et Vved. – Mogol-Tau, gorge of the Muzbek river, 800–900 m above sea level, shibliak, stony-pebble slope.

Generally accepted procedure [10,11] was used for determination of potential (PSP) and real (RSP) seed productivities and productivity coefficient (C) for a shoot in 50 mature generative individuals. Seeds were germinated in Petri dishes, 100 pieces each, in 3-fold replication at room temperature 20-24 °C for 10 days. Germination rate was expressed by the number of seeds germinated per day as a percentage of the total number of seeds germinated for 10 days. Data were treated with the use of STATISTICA 10.

3 Results and discussion

The species are similar in inflorescence structure – this is a capitate thyrse consisting of numerous opposite-arranged dichasia. A flower gynecium contains 4 ovules. *Ziziphora* flowers are entomophilous, their efflorescence is acropetal (from the flowers in the lower part of the inflorescence to the flowers in the upper part). In this regard seed ripening in inflorescences is stretched in time. *Z. interrupta* has a many-flowered inflorescence, the number of flowers in it is 1.5–2 times greater compared to the other species (Table 1), in this regard its PSP is also higher. Despite growth in different ecological-phytocoenotic conditions, the index of seed setting of the species does not differ and averages 1.4–1.6 seeds per fruit. Real seed productivity ranges from 31 to 71 seeds per inflorescence. However, the coefficient of productivity, an indicator of species reliability and wellbeing, does not exceed 40% in the species under study.

Table 1. Seed productivity of 4 *Ziziphora* species

Species	PSP		RSP		C, %
	Number of flowers in inflorescence	Number of ovules in inflorescence	Number of seeds in coenobium	Number of seeds per inflorescence	
Z. brevicalyx	20.9±1.07	83.7	1.5±0.5	31,4	37.5
Z. interrupta	44.9±2.66	179.5	1.6±0.7	71.8	40.0
Z. suffruticosa	32.7±1.52	130.9	1.6±0.6	52.3	39.9
Z. pamiroalaica	29.1±1.34	116.6	1.4±0.5	40.8	35.0

A fruit of the species under study – coenobium represents a fractional box in which there are 4 seeds. At disintegration of the box, erems in which there is one seed are formed. An erem coat grows together with that of a seed [12]. Seeds are ovate, oblong-ovate, glabrous or tubercular. Erem colour (seed) is light brown, greyish-brown, seldom brown. The largest seeds are noted in *Z. pamiroalaica,* mass of 1000 seeds is almost 1.5 times greater than in the other species. Distinctions are also observed in size, values of length and width of *Z. pamiroalaica* and *Z. brevicalyx* seeds differ significantly from those of the other two species (Table 2). Seeds in the experiment swell very well and germinate better at the light than in the dark. Swelling process does not depend practically on the temperature of the germination medium and completely comes to an end in 1-2,5 days since the moment of moistening. Good water permeability of seed coats of *Ziziphora* species indicates absence of physiological dormancy and excludes necessity of scarification. Laboratory germinating ability of seeds of three species amounts to 93–95%, it is somewhat lower in Z.

pamiroalaica – 88%. On the whole, germination percentage is very high (Fig. 1), which indicates superior quality of the species seed.

Table 2. Morphometric indices of *Ziziphora* species seeds

Species	Morphometric indices			
	Length, mm M ± m	Width, mm M ±m	Length/width	Mass (1000 seeds), mg
Z. brevicalyx	1.5±0.02	0.7±0.01	2.2±0.04	0.319±0.001
Z. interrupta	1.4±0.01	0.6±0.01	2.3±0.03	0.302±0.001
Z. suffruticosa	1.3±0.01	0.6±0.01	2.2±0.04	0.298±0.001
Z. pamiroalaica	1.7±0.02	0.7±0.01	2.4±0.03	0.402±0.002

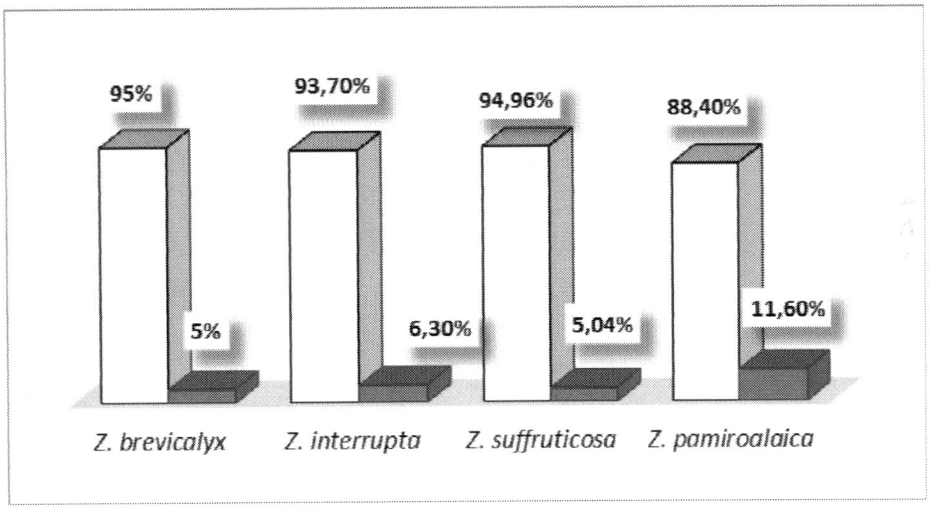

Fig. 1. Laboratory germinating ability of seeds of 4 *Ziziphora* species.

Designations: ☐ – germinated, ■ – died off.

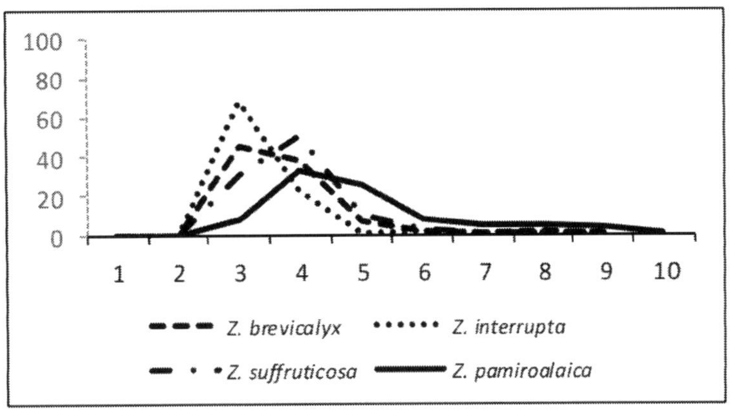

Fig. 2. Dynamics of seed germination of *Ziziphora* species at 20-24°C.

Designations: on the abscissa axis – days of germination, on the ordinate axis – number of germinated seeds.

Single-vertex curve of seed emergence rate (Fig. 2) with the maximum on the 3-d and 4-th days of the experiment is typical of the species. *Z. interrapta* seeds germinate faster all, 68 % of seeds germinated already on the 3-rd day, the main quantity of *Z. suffruticosa* and *Z. brevicalyx* seeds emerged for 2 days and accounted for 91 и 90 %, respectively. Extended germination during 7 days was noted only in *Z. pamiroalaica*. Such a character of seed germination is connected with growth conditions of the species. The first three species grew in the lower and middle belts of the mountains at an altitude of 800-2200 m above sea level in the arider conditions in the shibliak belt, and *Z. pamiroalaica* – at an altitude of 2800 m and higher above sea level in more humid conditions in the belt of semisavanna tall herbage.

High germinating ability and multiplicity of generative shoots compensate a low figure of the coefficient of inflorescence seed productivity. This makes it possible to maintain stability of populations in different montane belts of the Pamir-Alay. However, a short period from swelling to beginning of germination and high seed emergence rate at sharp fluctuations of weather conditions may adversely affect further existence and development of young plants.

The work was carried out within the framework of the projects of State Assignments ГР № 01011ТД065 (Tajikistan) and № АААА-А17-117012610053-9 (CSBG SB RAS, Russia).

References

1. C. Baskin, J. Baskin, Amer. J. Bot. **75**, 286-305 (1988)
2. E.A. Khodachek, Bot. J. **55**, 995-1010 (1970)
3. S.V. Bronnikova, Rastitelnye Resursy **35**, 43-48 (1999)
4. N.Yu. Kurochkina, Scien. Bul. of BSU, S. Nat. Sciences **98**, 183-187 (2011)
5. N.I. Gordeeva, Bul. of ASAU. **11**, 87-90 (2015)
6. T.V. Elisafenko, Contemporary Problems of Ecology **8**, 446–456 (2015)
7. P.N. Ovchinnikov, Flora of the Tajik SSR **1**, 9-20 (1957)
8. T.F. Kochkareva, Flora of the Tajik SSR **8**, 266-274 (1986)
9. A.L. Budantsev, E.E. Lesovskaya, *Useful wild plants of Russia* (St-Peterburg, 2001)
10. R.E. Levina, *Reproductive biology of seed plants (a review of the problem)* (Ul'yanovsk, 1981)
11. I.V. Vaynagiy, Bot. J. **59**, 826-831 (1974)
12. Z.T. Artushenko, *Atlas on descriptive morphology of plants* (Leningrad, 1986)

Spatial structure of the Kodar-Kalar orobiome botanical diversity on bioclimatic basis

Maxim V. Bocharnikov[1,*], and *Anton A.* Stas'ko[1]

[1]M.V. Lomonosov Moscow State University, 119991, Leninskie gory, 1, Moscow, Russia

Abstract. Bioclimatic substantiation of the vegetation spatial structure of the Kodar-Kalar orobiome on the basis of altitudinal vegetation divisions using the global climate model (BioClim) was carried out. Statistical analysis showed differences between altitudinal belts, sub-belts and also larch forests types on average annual temperature and average annual precipitation. The possibility of using the climate as a factor of differentiation of the vegetation cover at the regional level has been proved.

1 Introduction

The estimation of botanical diversity is necessary for its comparative geographical analysis, monitoring and development of environmental measures. The use the botanical-geographical units as a basis has advantages in diversity evaluation. The concept of biome diversity (the hierarchical structure was created by Walter and Brackle [1]) considers biomes as the basic regional unit for assessing biodiversity. It has the most optimal relationships with the abiotic environment, primarily with climatic conditions and landscape structure of the territory. Development of the concept has found on the map "Biomes of Russia" [2].

The biome concept allows estimating the biodiversity of heterogeneous in its spatial structure mountain areas on the basis of a complete altitudinal spectrum of vegetation. Each spectrum has a regional specificity expressed in types and groups of types of altitudinal zonality. In the mountains with altitudinal patterns of the vegetation cover structure and their integral expression in the types of vegetation belts at the regional level, the organization of the biota is distinguished by orobiomes. The spatial structure of the orobiome vegetation cover is reflected at different levels of the biota organization. The unity of orobiomes is based on the diversity of a complete altitudinal spectrum of vegetation in certain geographical conditions.

Orobiomes of regional level have a number of key criteria and indicators of biota and abiotic conditions. The specificity of orobiomes characterizes their biogeographic integrity as a combination of natural ecosystems developing under the influence of altitudinal regularities. These include the following: a complex of abiotic factors (primarily bioclimatic, geological, geomorphological, landscape) and the character of the altitudinal gradient of their quantitative indices; the level of floristic diversity (total species richness,

* Corresponding author: maxim-msu-bg@mail.ru

number of species per unit area) and the gradient of its variation along the altitudinal spectrum; the level of cenotic diversity (the number of basic and unique plant communities of altitudinal belt units); development of florocenogenetic complexes and their role in the vegetation cover; spatial patterns in altitudinal organization of the vegetation cover in connection with the specificity of prevailing processes. The main purpose of this investigation is evaluation of climatic factor in spatial structure of vegetation cover based on bioclimatic parameters.

2 Materials and methods

Selected criteria at the regional level should be used integrally for the whole orobiome. Bioclimatic characteristics have been selected as the main in the analysis of the Kodar-Kalar orobiome (the Northern Transbaikalia area). The general patterns of its structure in mountain conditions are reflected in the typology of altitudinal zonality [3] and in the level of biodiversity. The regional specifics are defined by the biome organization. Botanical-geographical mapping made a special contribution to revealing the structure of orobiome vegetation cover by the basis of altitudinal spectrum.

Information about the spatial structure of the Kodar-Kalar orobiome vegetation cover was taken from the vegetation map of the south of Eastern Siberia (s. 1: 1 500 000) [4]. Typological units related to the East Siberian mountain tundra, Altai-Tien-Shan alpine, Ural-South Siberian, Angarida and Beringia boreal complexes of plant communities were reflected in its legend in according to the geographic-genetic classification [5].

The global digital model BioClim (Hijmans et al., 2005) was taken as the climatic data. Raster layers (spatial resolution – 30 arc seconds) with continuous values of temperature and moisture supply parameters were used: the average annual temperature, the average summer temperature (warm quarter of the year), the average winter temperature (cold quarter of the year), the temperature amplitude (the difference between the maximum temperature of the warmest and lowest temperatures of the coldest months), the average annual precipitation, the average summer precipitation (warm quarter of the year), the amount of precipitation for the winter (the cold quarter of the year).

A quantitative interpretation of the regularities of vegetation distribution in connection with climatic conditions was carried out on the basis of a comparison of raster layers with climatic layers of BioClim. Altitudinal vegetation belts and types of larch (*Larix gmelinii* (Rupr.) Rupr.) forests as the key subdivisions of the mountain taiga belt were used in analysis. The main descriptive statistics for each of the typological subdivisions by the set of pixels were determined based on the spatial resolution of the climatic model: maximum and minimum values, arithmetic mean, its error and standard deviation from the mean. The veracity of the differences between the levels of temperature and moisture availability of vegetation of different altitudinal divisions and types of larch forests was estimated using Student's t-test.

3 Results

The parameters of the average annual temperature and the average annual precipitation diagnose altitudinal vegetation subdivisions for orobiome, changing proportionally to their position in the altitude spectrum (Table 1). The mountain taiga belt of larch forests is developed at the average annual temperatures from –5° to –10° C. Pine forests grow with the greatest temperature supply. In the interval of the average annual temperatures from –8° to –10° C the sub-tundra belt change the mountain taiga belt. In a wide range of temperatures (–9° –13° C) the vegetation cover of the mountain tundra and the upper tundra belts is formed. The amplitude of mean annual precipitation over altitudinal gradient is small. The lower sub-belt of the mountain taiga belt has the least moisture (about 400 mm of precipitation a year or less). It is associated with intermountain depressions with increased continental climate. Highlands have the highest moisture content (more than 500 mm of precipitation a year). In the warm period of the year, the increase in the amount of precipitation corresponds to the change of belts along the spectrum from the lower to the higher. In the cold period of the year, the dependence is nonlinear – the greatest amount of precipitation falls on the Siberian dwarf pine sub-belt and decreases up and down along the altitudinal spectrum.

Table 1. Bioclimatic indicators of the altitudinal vegetation belts and sub-belts of the Kodar-Kalar orobiome vegetation cover.

The altitudinal belts and sub-belts		I, II	III.1	III.2	IV.1	IV.2
Number of pixels		13135	13953	11256	45173	2642
Area, %		15	16	13	53	3
Bioclimatic indicators	Tyear	–10.9±2	-9.9±1.9	-9.4±1.5	–8.3±1.5	–6.9±1.7
	Tmax_year	9.7±1.9	10.9±1.9	11.6±1.3	12.6±1.7	13.9±1.7
	Prec_year	490±48	469±50	467±42	437±49	391±44
	Prec_max_year	286±32	264±37	246±26	235±33	215±37

The altitudinal belts and sub-belts: I – the upper tundra belt; II – the tundra belt; III – the sub-tundra belt: III.1 – the Siberian dwarf pine sub-belt, III.2 – the sparse forests sub-belt; IV – the mountain taiga belt: IV.1 – the larch forests sub-belt, IV.2 – the larch-pine and fir-pine-spruce forests sub-belt. The bioclimatic parameters (mean ± STD): mean annual temperature (Tyear); the average temperature of the warmest quarter of the year (Tmax_year); the average annual precipitation (Prec_year), the sum of the precipitation of the warmest quarter of the year (Prec_max_year).

The bioclimatic indicators are chosen also for estimation of spatial structure of larch forests types. The average annual temperature and the average annual precipitation are considered as the key bioclimatic indicators.

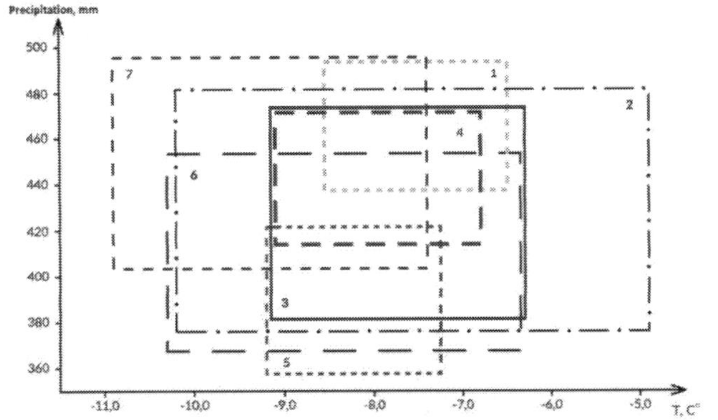

Fig. 1. Ordination of larch forests types according to the average annual temperature (T, C°) and annual precipitation (prec., mm).Larch forests types: 1 – *Rhododendron aureum*; 2 – *Rhododendron*

dauricum; 3 – *Vaccinium vitis-idaea*; 4 – *Duschekia fruticosa*; 5 – *Vaccinium uliginosum* and *Ledum palustre*; 6 – *Betula divaricata* and *B. exilis*; 7 – *Pinus pumila*.

4 Discussion

The altitudinal structure of Kodar-Kalar orobiome vegetation cover conform to the main bioclimatic gradients. Such patterns has been showed for vegetation belts and also typological units of forest vegetation [7, 8]. In this paper, the climatic conditions analysis of larch forests for the Kodar-Kalar orobiome (Angarida complex communities) as the basic vegetation for mountain taiga belt has been showed. They are formed in a wide range of conditions – from temperate cold and temperate humid ($\Sigma T>10°$ C – 1000-1400° C) with sparse larch forests on the boundary of the mountain taiga and sub-tundra belts to temperate warm and temperate humid (average annual precipitation is less than 400 mm) with communities in the Upper Angara and Muya-Kuanda depressions, where larch forests grow together with pine forests. A certain climatic range of larch forests of the orobiome in comparison with the larch (*Larix sibirica* Ledeb.) forests [6] is located in the range of same values of temperature supply, but with lower amount of precipitation (in mountains of Southern Siberia they grow at $\Sigma T>10°$ C to 1500° C and average annual precipitation to 600 mm).

Rhododendron (*Rhododendron aureum* Georgi) larch forests are common in the west and north-west parts of the orobiome, occupying the windward macro-slopes of the North Baikal and the Patom highlands. This explains the maximum annual precipitation among the larch forests types (460 mm). Rhododendron larch forests occupy a sufficiently high position in the altitudinal belt (700-900 m). The average annual temperature value of these forests does not drop below –7° C, which is the largest value among other types.

According to indication of climatic factor for units of phytosociological classification of Siberian and Far East vegetation, the larch forests are characterized by maximum continentality [9]. It is confirmed by the low values of the average annual temperature. In the first time the different amplitudes of temperature and moisture supply has been received and explained through altitudinal and geographic patterns for different types of larch forests based on geographic-genetic classification.

This research was supported by the grant of Russian Science Foundation (project № 17-77-10142).

References

1. H. Walter, S.-W. Breckle, *Okologishe Grundlagen in global sicht* (G. Fischer, Stuttgart, 1991)
2. *Map "Biomes of Russia" (S. 1: 7 500 000) in a series of nature maps for higher education*. Ed. G. N. Ogureeva (Financial and Organizational Consulting LLC, Moscow, 2015)
3. G.N. Ogureeva, M.V. Bocharnikov, Ecos.: Ecol. and Dyn. **1**, 52-81 (2017)
4. A.V. Belov, *Geobot. Map.* (Leningrad, 1973)
5. V.B. Sochava, *Geographical aspects of the Siberian taiga* (Nauka, Novosibirsk, 1980)
6. N.P. Polikarpov, N. M. Chebakova, D. I. Nazimova, *Climate and mountain forests of Southern Siberia* (Nauka, Novosibirsk, 1986)
7. E.O. Box, *Vegetation Science in Forestry* (Kluw. Acad. Publ., Dordtrecht, 1995)
8. S. Rivas-Martinez, D. Sanchez-Mata, M. Costa, Itin. Geobot. **12**, (1999)
9. Y. Nakamura, P.V. Krestov, A.M. Omelko, Phytocoen **37**, 433-470 (2007)

Forest vegetation assessment using geoinformation tools: a case of the Burla pine forest, Novosibirsk Region, Russia

Tatiyana S. Chernikova [1,*], *Yury* S. Otmakhov [1], and *Daria* D. Daibova [2]

[1]Central Siberian Botanical Garden, Siberian Branch of the Russian Academy of Sciences, Novosibirsk, 630090, Russia
[2]Siberian State University of Geosystems and Technologies, Novosibirsk, 630108, Russia

Abstract. The paper presents the vegetation thematic classification of the Burla banded pine forest carried on using "Canopus-V" remote sensing data and the supervised classification technique by a spectral angle mapper. Areas of selected elements have been assessed: 1. Pine forests, 2. Birch forests; 3. Meadows; 4. Anthropogenic objects (roads, etc.); 5. Agricultural lands; 6. Water objects. Sites of anthropogenic disturbed forests are identified according to remote sensing data. The results show that the data obtained in the classification by a spectral angle can be used to compile geobotanical maps, but due to low spectral resolution of Canopus-V satellite data, it is not always possible to classify individual objects validlys.

1 Introduction

Techniques of the Earth multispectral remote sensing are widely used to interpretate forest areas based on differences between their spectral properties recent decades. The data obtained allow mapping large areas, at the same time make it possible to draw preliminary conclusions about the nature of anthropogenic impact [1, 2, 3].

Pine forests have been attracting scholars' attention since the mid last century up to the present due to their large economic value [4, 5, 6, 7, 8, 9; 10, 11]. It was ascertained that intensive economic activity led to anthropogenic transformation of natural ecosystems. The share of pine forests was 92% in 1934, than it decreased to 83% by mass cuttings in 1963 [6, 12]. The banded forests were regarded as the raw material base permitting to receive high-quality wood, and their rational management was not taken into account. When the protective role of the banded forests was established and forest farms began actively restored, the area covered by the forest decreased from 288,000 to 85,000 hectares (as of 2015). However, planned logging operations are necessary, but should take into account biological features and environmental conditions of pine ecosystems [13, 14].

At the backdrop of the pine felling problems the issue of fires is acute as well. Many authors note not only an exceptionally high natural fire hazard of banded forests, but

* Corresponding author: hcernika@yandex.ru

temperature background rising, which lead to forest fires, their maximum was marked in 1990-1999. It has been established that fire careless handling is the main factor of fires [15, 16, 17, 18, 19].

Thus, anthropogenic impact on pine forests results natural community transformation, therefore complex monitoring of banded forests is necessary nowadays helping to solve a number of problems. Using the algorithms for decoding remote sensing data serves to identify quickly the damage foci and support operatively the security measure complex elaboration.

The main drawback of the most algorithms used is that each pixel in an image is compared with signatures of a test site (polygon) identified in the field and marked as a class. The supervised classification technique by a spectral angle mapper (SAM) allows comparing each pixel of an image with every finite element for each class and assigns a weighted value between 0 (low similarity) and 1 (high similarity). As a result, it is possible to obtain a classified image promptly and identify anthropogenic impact foci. The work objective is to assess SAM algorithm operation and determine the scale maximum possible for geobotanical mapping based on remote sensing Canopus-B data and field studies.

2 Materials and Methods

2.1 Study object

Pine banded forests are unique natural objects not only in scales of Siberia, but throughout Russia as well. They have come into being on alluvial sands along the ancient runoff valleys in steppe and forest-steppe zones. Pine forests bands (4-15 km width and 400 km length) stretch from northeast to southwest forming special climatic and soil conditions [20, 21].

Banded forests are situated in sharply continental climate regions with annual precipitation of 320 mm, significant temperature fluctuations up to 88 °C annual and 22 °C daily. These territories are characterized by hot dry winds, where pine forests act as a constraint [22]. Ridge-hilly relief causes the vegetation cover diversity in these forests. Due to such natural condition combination banded forests are unique, and preserve habitats of rare plant and animal species [23].

To achieve the objective, test sites were selected in the Burla forest band. It is located at Altai Territory mainly and its part enters southern Novosibirsk region. The band begins 35 km north of Kamen'-on-Ob city at the Ob reservoir shore and stretches from northeast (N54°14'05.78 "E81°39'59.60") to southwest (N53°46'12.10 "E80° 07'37.40") for 114 km with 8-10 km width. The bottom of the forested valley is 150-180 m altitude above sea level. The Priobskoe Plateau surrounding parts rise above it by 60-100 m. The surface sediments are sands underlying by clay shallow layers and lenses, which are waterproof deposits [24]].

2.2 Earth Remote Sensing Space System

Russian satellite Canopus-V remote sensing data have been used (surveying June 27, 2017). This space complex is designed to monitor man-made and natural emergencies, to detect forest fire foci, and can be used to map vegetation at a scale of 1 : 50,000 as well. It is known that for mapping at 1 : 25,000 scale with a minimum allowable index of map content, space images with a linear resolution not less than 2.5 m should be used. The obtained data have the following resolutions: panchromatic – 2.1 m gt; multispectral – 10.5 m. Spectral ranges are panchromatic (P) – 0.52-0.85 μm; blue (B) – 0.46-0.51 μm; green

(G) – 0.54-0.6 μm; red (R) – 0.63-0.66 μm; near infrared (Nir) – 0.75-0.86 μm [25]. Applying a Pan-sharpening function, proper data for the vegetation visual interperting, and relevant materials for mapping at 1 : 25,000 scale can be got.

2.3 Geoinformatiom methods

The applied image processing technology allows identifying images in accordance with their spectral signatures in a semi-automatic mode (supervised classification). There are several types of classification algorithms, but the general purpose is to compile the vegetation thematic map of. At the first stage, a composite raster has been created by five bands: P, R, G, B, Nir. An image obtained by four channels (R, G, B, Nir) is used for classifying and creating spectrograms. At the second stage, a shapefile of the regions of interest (ROI) and signatures has been generated. Areas have been singled out by polygon capture of pixels. The test sites determined by routing geobotanical study have been chosen: 1. Pine forests, 2. Birch forests; 3. Meadows; 4. Anthropogenic objects (roads, etc.); 5. Agricultural lands; 6. Water objects. The supervised classification algorithm by SAM has been used for classifying. Preliminary processing of Canopus-V remote sensing data, classification and final maps arranging have been carried on with free cross-platform geoinformation system QGIS [26] applying a semi-automatic classification module Plugin [27, 28], projection code is EPSG: 32644.

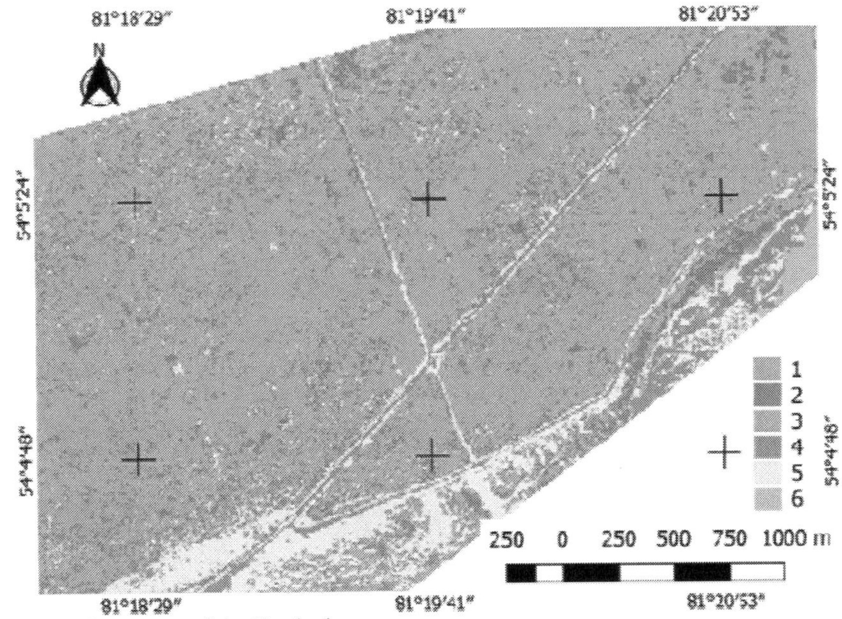

Fig. 1. A map fragment of the Burla forest.

3 Results

As a result of applying the supervised classification technique by SAM based at test sites of five classes set at an assigning stage, the areas of different vegetation units identified (Fig. 1). When classifying vegetation, the result accuracy depends directly on the remote sensing data resolution. In our case, the resolution of Canopus-V remote sensing data is 10×10 m, so if the object (roads, local felling, etc.) size in an image cell is less than the image resolution, the pixel reflection coefficient is a linear combination of reflection coefficients of all surface types entering this pixel, which leads to some misrepresentation of the classification results. Due to the similarity of spectral signatures of disturbed territories with a size less than remote sensing data resolution with the signatures of the open soil (road), the classifier could not uniquely distinguish them, therefore the swamped lake coastal areas fell into the group of anthropogenic objects (4). The same happened with agricultural lands (5): some of anthropogenic objects (4) were classified as croplands (5). Therefore, it was decided to allocate all the disturbed territories to a separate class. This does not play a significant role in the area estimation of vegetation types, as the most of the studied objects are slightly impacted by anthropogenic effect. When assessing the accuracy of SAM technique applied, high values have been marked and the same common accuracy is shown. In general, the use of classification by SAM allows determining the areas of the identified vegetation types the most accurate.

4 Discussion

This work on the vegetation classification makes it possible to assess the suitability of the remote sensing data analysis for monitoring the Burla banded pine forest. Using remote sensing data from the Russian satellite Canopus-V and semi-automatic algorithms permits to obtain initial materials for interpretring vegetation and further mapping at 1: 25,000 scale. The classification finalization enables to identify anthropogenic objects (cut and burned sites, roads, etc.). As noted above, anthropogenic objects of small sizes have some errors in defining their class and areas, but still with considerable accuracy it is possible to reveal disturbed territories. The natural vegetation (a vegetative component) analysis of the studied region is of major importance. Despite the revealed potential problems of the classification process, the maps obtained can be an additional predictor for geoinformation modeling, and serve the purposes of analyzing anthropogenic and natural changes in forest ecosystems.

The project supported by RFBR (Grant No 16-05-00908 A) and research was carried on in frameworks of base projects of the Central Siberian Botanical Garden, the Siberian Branch of the Russian Academy of Sciences (Project no. VI.52.1).

References

1. Zh. Li, D. Limin, G. Shao, X. Dong, W. Hui, B. Jianwei, Science in China. *E*. **49**, 167-176 (2006)
2. L. Altun, E.Z. Baskent, A. Gunlu, A.I. Kadiogullari, Environ Monit Assess. **137**:149-161 (2008)
3. S. Arekhi, A.A. Jafarzadeh, Arab. J Geosci. **7**:1073-1085 (2014) DOI 10.1007/s12517-012-0785-1
4. P.L. Gorchakovsky, Bot J. **34**, 525-538 (1949). (In Russ.)
5. G.G. Pavlova, Vegetation of steppe and forest-steppe zones in West Siberia **6**, 131-162 (1963) (In Russ.)
6. I.V. Taran, *West Siberian pine forests* (Nauka, Novosibirsk, 1973) (In Russ.)
7. I.A. Khrustalyova, Bulletin of Altai State Agricultural University **8**, 32-35 (2008) (In Russ.)
8. I.A. Khrustalyova, Plant Life of Asian Russia **2**, 54-65 (2009). (In Russ.)
9. T.A. Teryokhina, T.M. Kopytina, P.A. Kosachev, N.V. Elesova, N.V. Ovharova, Proc. South. Siber. Bot. Garden **12**, 54-65 (2008) (In Russ.)
10. N.N. Lashchinsky, A.Yu. Korolyuk, N.V. Lashchinskaya, Plant Life of Asian Russia **2**, 63-84 (2017) (In Russ.)

11. N.N. Lashchinsky, A.Yu. Korolyuk, M.P. Tishchenko, N.V. Lashchinskaya. Asian Russia Vegetative Kingdom **1**, 57-81 (2018). (In Russ.)
12. A.E. Osipenko, E.M. Ananiev, A.A. Gof, V.V. Savin, D.A. Shubin, News of the Orenburg State Agrarian university **6**, 98-101 (2017). (In Russ.)
13. A.A. Maslenko, M.I. Semenov, V.A. Elizarov, Bulletin of Altai State Agricultural University **12**, 54-57 (2012). (In Russ.)
14. E.G. Paramonov, Bulletin of Altai State Agricultural University **8**, 55-59 (2014). (In Russ.)
15. L.P. Barannik, V.I. Zablotsky, Bulletin of Altai State Agricultural University **3**, 61-64 (1999). (In Russ.)
16. E.G. Paramonov, Ya.N. Ishutin, *Large forest fires in Altai Region* (Del'ta, Barnaul, 1999). (In Russ.)
17. E.Yu. Neverova, A.A. Malinovskikh, Bulletin of Altai State Agricultural University **12**, 58–61 (2012) (In Russ.)
18. I.M. Frolenkov, O.M. Frolenkov, World Science, Culture, Education **1**, 394-397 (2014) (In Russ.)
19. E.G. Paramonov, I.D. Rybkina, Sustainable Forestry **3**, 33-39 (2017). (In Russ.)
20. L.N. Gribanov, *Banded pine forests in Altai Region and Kazakhstan* (Sel'khozgiz, Moscow, 1954). (In Russ.)
21. G.V. Zanin, Natural zoning of the Altai territory **1**, 62-98 (1958). (In Russ.)
22. A.P. Slyadnev, Ya.P. Fel'dman, Natural zoning of the Altai territory **1**, 9-61 (1958) (In Russ.)
23. *The red book of the Altai Region: rare and endangered plant species*. Vol. 1 (Altai Univ. Publ., Barnaul, 2006). (In Russ.)
24. I.P. Gerasimov, Materials to the geomorphology of the Kulunda steppe **10**, 9-33 (1935) (In Russ.)
25. A.V. Vladimirov, R.S. Salikhov, N.A. Senik, S.A. Zolotoi, Problems Electromechanics **105**, 49-57 (2008) (In Russ.)
26. QGIS Development Team. QGIS geographic information system. Open Source Geospatial Foundation (OSGeo). URL: https://qgis.org/ru/site/index.html (accessed 30.05.2018).
27. L. Congedo, M. Munafò, S. Macchi, *Investigating the relationship between land cover and vulnerability to climate change in Dar es Salaam* (Sapienza Univ., Rome, 2013).
28. L. Congedo, Semi-automatic classification Plugin. Documentation. Release 6.1.0.1, 2018, URL: https://media.readthedocs.org/pdf/semiautomaticclassificationmanual/latest/semiautomaticclassificationmanual.pdf (accessed 30.05.2018)

BIO Web of Conferences **11**, 00009 (2018) https://doi.org/10.1051/bioconf/20181100009
Prospects of Development and Challenges of Modern Botany

The morphometric characters variability analysis of *Tulipa uniflora* (Liliaceae) in the Angara Region (Irkutsk Oblast, Russia)

Olga Chernysheva[1,*], *Yuriy* Bukin[2], and *Denis* Krivenko[1,3]

[1]Siberian Institute of Plant Physiology and Biochemistry SB RAS, 664033 Irkutsk, Russia
[2]Limnological Institute SB RAS, 664033 Irkutsk, Russia
[3]Federal State Budgetary Institution Western Baikal Protected Areas, 664050 Irkutsk, Russia

Abstract. The intra- and interpopulation variability of six populations of *Tulipa uniflora* (L.) Besser ex Baker at the northern limit of distribution in the Angara steppes over 12 morphometric parameters was studied by the principal components method and single-factor analysis of variance. It is established that the vectors characterizing the gradients of the change in the morphometric parameters are divided into two groups: vegetative and generative, which are among themselves in direct dependence. Based on the single-factor analysis of variance results for all morphometric parameters, a significant difference is observed between populations of *T. uniflora* in the Angara Region. On the one hand, this is a consequence of anthropogenic impact on the population, and on the other – the relic nature of the species.

1 Introduction

Wild tulips are famous ornamental plants. They are the ancestors of cultural species and an inexhaustible source for breeding new varieties. They have an invaluable advantage over cultural plants: an exceptionally diverse color of the perianth, they are immune to the variegation virus [1]. Natural types of tulips need a wider introduction to culture, including gardening in the territory of Siberia. The study of the morphometric characters variability of the species of the genus *Tulipa* L. is of great importance in assessing the state of natural populations of species. The survey of as many tulip sites as possible allows revealing key indicating morphometric characters and ecological and biological optimums typical for each species. The presented study is aimed at the analysis of the variability of *T. uniflora* morphometric characters, including the analysis under the influence of anthropogenic impact on populations at the northern border of growth in the steppes of the Angara Region.

2 Materials and Methods

* Corresponding author: helga8408@mail.ru

T. uniflora is mainly a Central Asian steppe species. In South Siberia the species is registered in the Central Altai, Tuva, Ubsunur, Khakass, Angara Region and Daurian steppes [2, 3]. At the northern border of growth *T. uniflora* belongs to the category of rare and endangered plant species. The species is a tertiary relic of the Miocene-Pliocene flora. The life cycle of the species is in disharmony with the existing natural conditions. *T. uniflora* blossoms in late April–early May. Early flowering is a necessary phenomenon in Central Asia where it is determined by the aridity of summer. Since in South Siberia drought usually occurs in May–June and precipitation falls in July–August, the mentioned features of the life cycle are explained only by historical causes.

The studies were carried out in 2017 in Ekhirit-Bulagatsky, Ust-Udinsky, Osinsky, Bokhansky Districts of the Irkutsk Oblast (Table 1). In each population up to 12 morphometric characters were measured in 15–25 random specimens which were in a generative ontogenetic state.

Table 1. The studied populations of *Tulipa uniflora* in the Irkutsk Oblast.

Location – Symbol designation	N / E	Stage of grazing digression [8]
Bokhansky District, Khandagai – *H*	53°11′12.5″ / 103°37′01.7″	3–4
Bokhansky District, Bohan – *B*	53°09′40.8″ / 103°48′02.9″	9
Osinsky District, Lenino – *L*	53°41′26.5″ / 103°45′23.8″	6–7
Osinsky District, Obusa – *O*	53°43′18.2″ / 103°49′15.0″	5
Ust-Udinsky District, Igzhey – *I*	54°03′25.7″ / 103°62′25.6″	6–7
Ekhirit-Bulagatsky District, Gahany – *G*	53°02′05.1″ / 104°52′46.0″	3–4

Due to the fact that *T. uniflora* has a short blossoming period, it was not possible to measure some morphometric characters of the flower in the populations *B*, *H*, *I*, *L*; therefore the original data table had missing data. The restoration of the missing data was carried out using the multidimensional linear regression according to the recommendations of W.R. Myers [4] and K.J. Lee & J.B. Carlin [5].

Before the statistical analysis the array of morphometric data was normalized with the help of a standard metric with the purpose to bring all the considered morphometric parameters to a single scale of measurement and to balance their contribution to the variability of the data array. To analyze and visualize the distribution of the specimens under study for a set of morphometric characters, one of the methods of multidimensional statistics (the method of principal components) was used [6].

To assess the measure of discrepancy in the morphometric characters in the populations studied, we used the single-factor analysis of variance with the Fisher criterion ($\alpha = 0.05$) [7]. The measure of discrepancy (H) between populations according to the test feature was calculated by the formula: $H = 1 - 1/F$, where F is the value of the Fisher criterion. H takes a value from 0 to 1; the closer it is to 1, the more the populations differ. The nature and extent of anthropogenic impact in the populations were estimated by the pasture degradation scale of L.G. Ramensky and others [8]. All the statistical calculations and the construction of the diagram were carried out in a shareware statistical environment for programming purposes R.

3 Results

The scatter diagram characterizing the specimens under study in the space of the first two principal components is shown in the figure. Point clouds which combine the specimens in the populations have a different overlapping. The greater the overlapping of clouds is, the more similar the populations are among themselves according to the morphometric characters under analysis. The most similar populations are ***B, G, H, L*** and ***O***. The specimens from the population ***I*** are the most different from the specimens of the other populations. The specimens from the populations ***G*** and ***I*** differ significantly.

The vectors characterizing the gradients of the change in the morphometric characters on the scatter diagram (Fig.) are divided into two groups. The co-directional vectors of the parameters of the length and width of the prophyll and the secondary leaf, the stamen height and the generative shoot height are associated in the first group. The vectors of the parameters of the length and width of the inner and outer perianth and the pistil height are associated in the second group. These parameters are related to each other by the direct dependence in both groups.

With an increase in the value of one of the analyzed morphometric characters, the other parameters also increase as a rule. The two groups of vectors are almost perpendicular to each other. That indicates that there is no correlation between these groups of features.

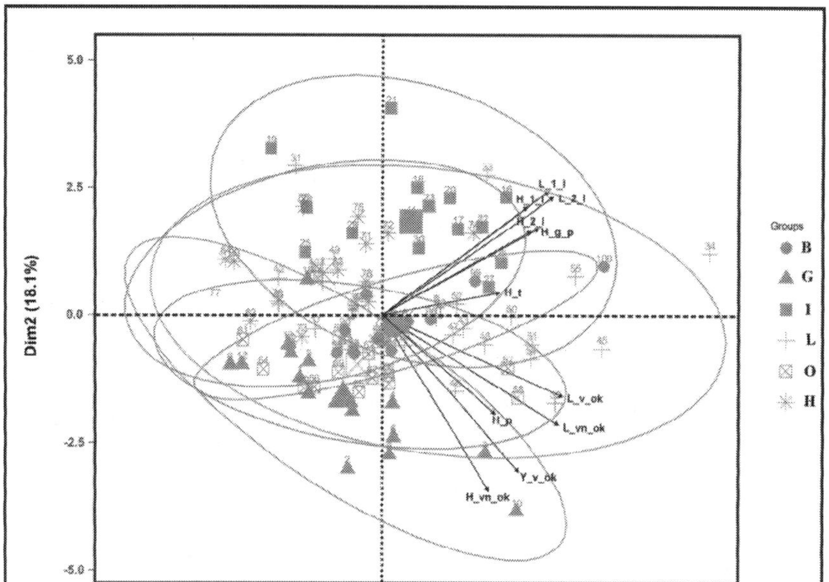

Fig. Scatter diagram characterizing the studied specimens of *Tulipa uniflora* in the space of the first two principal components. Ellipses outline the point clouds, the concentration of specimens from the certain populations. The vectors show the gradient direction of the change in the morphometric characters under study. The designation of populations is given in Table 1, the designation of the features is given in Table 2.

The results of the single-factor analysis of variance are shown in Table 2. There is a significant difference between the populations for all the features. The greatest difference is observed in the stamen height (H = 0.907). Three features are least distinguished: the length of the inner and outer perianth and the pistil height (H < 0.7). For the other features the value of H exceeds 0.8; that indicates a fairly significant level of difference between the populations.

Table 3. Results by a single-factor variance analysis of *Tulipa uniflora*.

Name of morphometric character – Symbol designation	F	H	P_value
Height of generative shoot – H_g_p	7.290	0.862	0.000
Length of first leaf – L_1_l	9.629	0.896	0.000
Width of first leaf – H_1_l	7.167	0.860	0.000
Length of second leaf – L_2_l	9.299	0.890	0.000
Width of second leaf – H_2_l	6.481	0.845	0.000
Length of outer perianth – L_v_ok	3.263	0.693	0.009
Width of outer perianth – Y_v_ok	9.862	0.898	0.000
Length of inner perianth – L_vn_ok	3.434	0.708	0.006
Width of inner perianth – H_vn_ok	9.103	0.890	0.000
Height of stamen – H_t	10.800	0.907	0.000
Height of pestle – H_p	3.019	0.668	0.014

4 Discussion

Morphological differentiation of the specimens of the population *I* from the specimens of the populations *B, G, H, L* and *O* can be explained by the geographic remoteness of the first from the others by 60–160 km and by the significant anthropogenic impact on it (Table 1). The habitat of *T. uniflora* of the population *I* is confined to the gentle slope (< 30°); that is a convenient place for grazing cattle and small ruminants; the proximity of the ferry and motorway makes the slope accessible to people. Soil erosion was noted here as a consequence of the anthropogenic impact. The specimens from the populations *I* and *G* differ significantly according to their morphometric characters. This is due to the remoteness of the populations from each other by 160 km, the natural isolation of the population *G*, including the populations *B, H,* L and *O*, the Kuda River and forestlands. The latter circumstance is an obstacle for cross-pollination between the specimens of the population *G* with the other populations. It should be noted that the habitat of the population *G*, unlike the population *I*, is undisturbed (Table 1) despite the proximity of the locality. However, the slope where *T. uniflora* dwells is much steeper (~ 50°) which makes it difficult to graze livestock.

To dividing vectors into generative and vegetative groups is consistent with the data obtained by L.V. Mukhametshina and others [9] for the South Ural populations *T. biebersteiniana* Schult. & Schult.f., *T. patens* C.Agardh ex Schult. & Schult.f. and *T. riparia* Knjaz., Kulikov & E.G.Philippov.

For the *T. uniflora* populations in the Angara Region the feature of the stamen height is knocked out from the general direction of the vectors of generative features. According to the results of the single-factor analysis of variance according to the Fisher criterion, the greatest difference among the populations is observed on this feature (Table 2). This is probably related to the fact that in some populations the phasing out of the stage of tulip blossoming was observed.

The low variability of the morphometric characters of the generative sphere in comparison with the features of the vegetative sphere in the populations is quite natural, since the flower has an exceptional role as a special morphological structure associated with the fact that it completely combines all the processes of agamic and sexual reproduction. Excluding microclimatic and relief-forming factors, the mentioned features of the variability of the morphometric characters in the specimens of *T. uniflora* from the populations of the Angara Region can also be explained by historical causes, namely the prolonged genetic isolation of the peripherical populations from the main part of the area of species.

The reported study was funded by RFBR according to the research project № 18-34-00551.

References

1. Z. P. Bochantseva, *Tulips. Morphology, cytology and biology.* (Publishing House of the Academy of Sciences of the Uzbek SSR, Tashkent, 1962)
2. N.V. Vlasova, *Tulipa* L. *Flora Sibiri*, 4 (Nauka, Novosibirsk, 1987)
3. G.A. Peshkova, *Florogeneticheskii analiz stepnoi flory gor Yuzhnoi Sibiri* (Science Publishers, Inc., Novosibirsk, 2001)
4. W.R. Myers, Drug. Inf. J. **34**, 525–533 (2000)
5. K.J. Lee, J.B. Carlin, Am. J. Epidemiol. **171**, 624–632 (2010)
6. S.A. Aivazyan, V.M. Buchstaber, E. S. Enyukov, et al., *Prikladnaya statistika. Klassifikatsiya i snizheniye razmernosti* (Finance and Statistics, Moscow, 1989)
7. J.M. Chambers, A. Freeny, R.M. Heiberger, *Analysis of variance; designed experiments* (Statistical Models in S, Cole, 1992)
8. L.G. Ramenskii, I.A. Tsatsenkin, O.N. Chizhikov, et al., *Ekologicheskaya otsenka kormovykh ugodii po rastitelnomu pokrovu* (Sel'khozgiz, Moscow, 1956)
9. L.V. Mukhameditdishina, E.Z. Mullabaeva, M.M. Ishmuratova, *Izvestiya Samarskogo nauchnogo tsentra RAN* **16**, 1650–1653 (2014)

BIO Web of Conferences **11**, 00010 (2018) https://doi.org/10.1051/bioconf/20181100010
Prospects of Development and Challenges of Modern Botany

Plant characteristics determination in sand pits near Moscow

Alaa Dabbagh[1,*]

[1]Faculty of Ecology, PFUR, post-graduate student, Miklukho-Maklaya street 6, 117198, Moscow, Russia.

Abstract. Sand pits are man-made landforms, where vegetation self-restoration becomes very difficult. An in-depth study of sand pits' flora and vegetation must be done in order to restore the complete canopy on sand substrate, which prevents the movement of slopes. In Zvenigorod and Dzerzhinsk sand pits, aged about 70 years, we have studied the slopes with northern and southern exposure for 3 years. The article shows that all the 131 species which we have found in the surveyed area were species of vascular plants. Projective cover degree made up 20-30% which, in general, is a typical characteristic of sandy terrain. Hemicryptophytes dominate the spectrum of life forms by K. Raunkiaer in our research. According to the classification of life forms by I. G. Serebryakov, sand slopes are dominated by herbaceous perennial polycarpics, among which long and short rhizome plants and tap rooted plants are in majority. A significant proportion of tap rooted plants is due to their high degree of adaptation to strong light conditions and resistance to drought. Long- and short rhizome plants tend to loose substrates and soft soils.

1 Introduction

Sand, the main reserves of which are confined to the moraine-glacial deposits of the Quaternary period, is extracted by the open method. After the completion of the work, the sand pits must undergo reclamation: with mechanical profiling and subsequent revegetation, but in most cases the vegetation of the sand pits is left to self-repair [1]. For more rapid reclamation and the prospects for further use of disturbed lands, it is important to study the ecological features of plant communities' growth in sand pits.

The objective of this research is to examine ecological characteristics of sand pits vegetation (in two sand pits near Moscow). The results of our study can be used to plan the processes of stabilizing sandy slopes or reclamation of sand pits. The obtained data supplement the available data on the regularities of the formation of plant communities. Quarry studies are of great importance in terms of developing recommendations for the restoration of vegetation or the use of disturbed territories for economic purposes.

2 Materials and methods

* Corresponding author: alaa_dabbagh@hotmail.com

Materials for the work were collected in the summer of (2015, 2016, and 2017) during field research in two sand pits of Moscow region (Zvenigorod and Dzerzhinsky). For future comparison, the observations were made in each sand pit on the slopes of northern and southern exposure. To study species diversity geo-botanical description is carried out using the standard method in the area of 5x5 meters. Basic ecological and bio-morphological characteristics for each species of plant were settled. Life forms are defined by the system K. Raunkiaer and I.G Serebryakov [2.3]. Ecological characteristic assessment of habitats was carried out using G. Ellenberg ecological scales [4]. Ecological-cenotic groups were determined according to M.V. Kazakova [5]. Plant names are given by I.A. Gubanov [6].

3 Results and Discussion

3.1 Species diversity of vegetation of sand pits

Determination of the species diversity of vegetation in the study areas is fundamental for the formation of a strategy for restoring the vegetation cover on a sandy substrate: only by inventorying the species can one go on to selecting the range of plants for further work. In both sand pits (Zvenigorodsky and Dzerzhinsky) vegetation was rarefied. Only 131 species were found in the surveyed area of the sand pits during the three years of our research. Projective cover degree made up 20-30% which, in general, is a typical characteristic of sandy terrain [7]. In this case the number of species is not only determined by the smooth running development of the landscape (taking into account the fact that the interference was rude and had a man-made factors), but by other factors such as steep slopes, light, moisture, and richness of the soil.

3.2 Spectrum of life forms of plants in sand pits according to K. Raunkiaer

Hemicryptophytes are typical life form, prevailing in all plant communities of central Russia. In arid zones, terophyte should dominate according to K. Raunkiaer. (Fig.1).

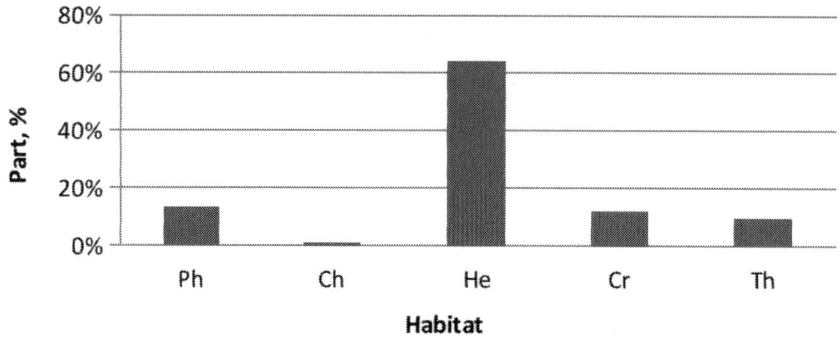

Fig. 1. Spectrum of life forms according to the system of K. Raunkiaer on different habitats. Notation keys: Ph-phanerophytes, Ch-chamaephytes, He-hemicryptophytes, Cr-Cryptophyte (geophytes), Th-therophytes

Hemicryptophytes dominate on the northern and the southern slopes of sand pits which is typical of central Russia vegetation. Compared to the desert regions terophytes low share can be noted. On the contrary, the ratio (the proportion) of geophytes is low compared to the moderate area. The predominance of hemicryptophytes may be noted on the northern slopes (with a set of environmental factors more zonal) compared to more hot and dry southern slopes, reflects the typical climatic conditions of the temperate zone and the originality of the substrate on which the plants grow. It may be noted that Chamaephyte with a low share. Such differences, most likely, are not associated with the exposure of the slopes but with the species diversity of vegetation that spontaneously develops in each sand pit.

3.3 Spectrum of life forms of plants in sand pits according to IG Serebryakov

I.G Serebryakov more elaborated and detailed, classification of life forms compared to those of K. Raunkiaer allows for detailed analysis of the environmental factors impact on plant community.

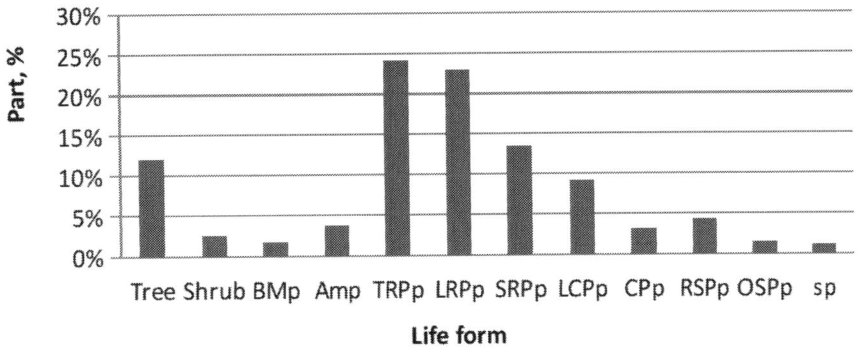

Fig.2. the spectrum of life forms by I. G. Serebryakov on different habitats. Notation keys: BMp – biennial monocarpic; Amp – annual monocarpic; TRPp – tap root polycarpic; LRPp – long rhizome polycarpic; SRPp – short rhizome polycarpic; LCPp – loose cespitose polycarpic; CPp – creeping polycarpic; RSPp – root sucker polycarpic; OSPp – overground stoloniferous polycarpic ; Sp – semishrub

According to life forms classification by I. G. Serebryakov (Fig. 2), sand slopes are dominated by herbaceous polycarpics, among which long- and short rhizome plants and tap rooted plants are in majority. A significant proportion of tap rooted plants is due to their high degree of adaptation to strong light conditions and resistance to drought [8]. Long- and short rhizome plants tend to loose substrates and soft soils, although the latter are more adaptive to the growing conditions. The presence of tap rooted plant due to their fitness to habitat high insolation and resistance to drought [9].

4 Conclusions

A study of the vegetation of the sand pits of the near Moscow region made it possible to identify and confirm the characteristic features of the formation of cenoses of these difficult habitats in ecological conditions. In particular:

1. The vegetative cover of sand pits is rarefied (in comparison with typical conditions of the humid taiga zone), the projective cover degree made up 20-30%, which is typical for sandy landscapes.
2. In the spectrum of life forms according to K. Raunkier, hemicryptophytes predominate for sandy pits of the Moscow region, which is typical of central Russia vegetation
3. In the spectrum of life forms by I.G. Serebryakov for sand pits are dominated by herbaceous polycarpics, among them dominant are long- and short rhizome plants and tap rooted plants

References

1. N.A. Berezina, N.B. Afanas'eva, *Ekologija rastenij* (Academa, Moscow, 2009)

2. AM. Aleynikova, EA. Vanisov, E.Yu. Vasilyeva, *Izuchenie prerodnikh ikosystems* (RUDN, Moscow, 2015)

3. Yu.E. Alekseev, P.Yu. Zhmylov, E.A. Karpukhina. *Rastenya gorada Dubna* (IUNSAM, Moscow, 2014)

4. P.Ju. Zhmyljov, A.P. Zhmyljova, E.A. Karpukhina, *Opredelenie uslovij sredy po rastitel'nosti* (Energija, Moscow, 2013)

5. M.V. Kazakova, *Flora Rjazanskoj oblasti* (Russkoe slovo, Rjazan, 2004)

6. I.A. Gubanov, K.V. Kiseleva, V.S. Novikov, V.N. Tihomirov. *Opredelitel' sosudistyh rastenij centra evropejskoj Rossii* (Argus, Moscow, 1995)

7. B.A. Bykov, *Osnovnye ekobiomorfy pustynnyh rastenij Turanskoj nizmennosti* (Kazakh Science. CCP., Almaty, 1987)

8. E.V. Abakumov, Pochvovedenie, E**8**, 955-963 (2008)

9. H.T. Gajrabekov, S.B. Macaev, M.V. Geroeva, Vestnik KrasGAU. E**4**, 128-133 (2012)

BIO Web of Conferences **11**, 00011 (2018) https://doi.org/10.1051/bioconf/20181100011
Prospects of Development and Challenges of Modern Botany

Identification of artificial interspecific hybrids in the genus *Elymus* (Poaceae) by using ISSR-markers

Maria Emtseva*, and *Alexander* Agafonov

Central Siberian Botanical Garden CSBG, Zolotodolinskaya st., 101, Novosibirsk, Russia, 630090

Abstract. Reproductive relationships between species of the genus *Elymus* (Poaceae) from mixed populations of the Southern Urals were studied. The true hybrids were identified by using ISSR-markers in comparison with parental biotypes. The presence in the true hybrids of species-specific ISSR-bands from both parental forms is provided by high level of homozygosity of species in genus *Elymus*, as a result of predominant self-pollination.

1 Introduction

The genus *Elymus* L. consists of polyploid perennial species of the tribe Triticeae Dum. (Poaceae). Taxa of the genus have haplomes “St”, “H”, “Y”, “P”, “W”, which originated from other genera of tribe Triticeae in different combinations, but “St” haplome presents in all *Elymus* species [1]. “St” haplome descended from *Pseudoroegneria* genus, “H” haplome – from *Hordeum* genus, “P” haplome – from *Agropyron* genus, “W” haplome – from *Australopyrum* Löve; the origination of “Y” haplome was not determined exactly yet. The genome constitution of *E. mutabilis* (Drob.) Tzvel. and *E. caninus* (L.) L. was identified as StStHH (2n=4x=28) [1]. In *E. uralensis* (Nevski) Tzvelev and *E. viridiglumis* (Nevski) Czer. the chromosome number 2n=28 was determined too [2], but genome formula was not identified conclusively.

Previously by using six ISSR-primers we showed the existence of genetic distances between biotypes of *E. uralensis*, *E. viridiglumis*, *E. mutabilis* and *E. caninus* growing on the closely located areas of Southern Urals, and significant distinctions between groups of Ural and Siberian species were revealed (Fig. 1).

* Corresponding author: emtsevamv@yandex.ru

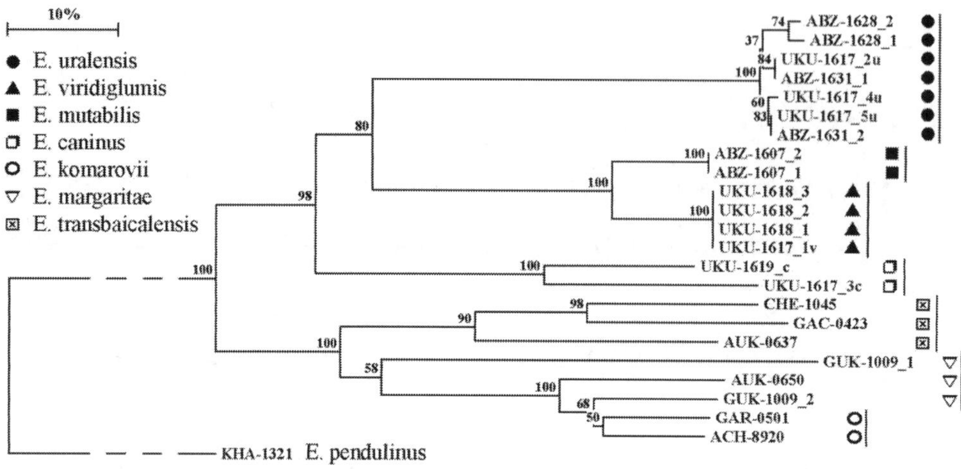

Fig. 1. Neighbour-Joining consensus tree with bootstrap support values based on results of the use of six ISSR-primers.

The multiple reticulate microevolution (which includes periodic hybridization and introgression), as a result of these species growing together during some generations, was suggested. For the study of reproductive compatibility of these species we created a series of hybrid plants between selective biotypes from mixed populations. The purpose of this work was identification of hybrid F_1 plants between *E. uralensis*, *E. viridiglumis*, *E. mutabilis* and *E. caninus* based on their specificity on ISSR-markers.

2 Materials and Methods

F_1 hybrids were created on the experimental plot of CSBG SB RAS (summer vegetation of 2017) by using the previously developed method [3]. These F_1 plants were grown in climate chamber of CSBG SB RAS. DNA was isolated from dried leaves by using kit for isolation of genomic DNA NucleoSpin Plant II (Macherey-Nagel, Germany). The PCR was set with ISSR-primers HB12 ($(CAC)_3GC$), M2 ($(AC)_8(C/T)G$) and M11 ($(CA)_6(G/A)$). The 15 μl mix of PCR contained 10 mM primer, 25 mM $MgCl_2$, 5 mM deoxynucleotides, 1x PCR-buffer, 0,9 units of Taq DNA polymerase and 2 μl of template DNA. PCR was carried out in C 1000 Thermal Cycler (BioRad Laboratories, USA). Annealing temperature for HB12 and M11 primers was 51°C, for M2 primer – 56°C. The obtained PCR-products were stained by SYBR-Green dye and separated by 1,5% agarose gel electrophoresis in 1x TBE buffer. The size of ISSR-fragments was determined according to 1 kb DNA ladder. Results of PCR were visualized by Gel Doc XR+ gel documentation system (BioRad Laboratories, USA).

3 Results and Discussion

We have obtained 20 plants in 7 cross combinations among *E. uralensis*, *E. viridiglumis*, *E. mutabilis* and *E. caninus*. Results of ISSR-analyses of assumed hybrids are shown in Fig. 2 in comparison with parental biotypes.

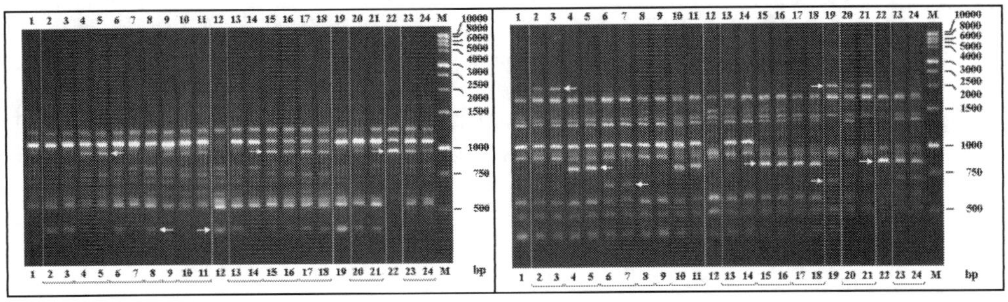

Fig. 2. ISSR-analysis of *E. uralensis*, *E. viridiglumis*, *E. mutabilis and E. caninus* and of hybrids between them with primers HB12 (left) and M11 (right). **1.** *E. uralensis* UKU-1617-2u; **2, 3.** uralen ABZ-1634 × mutab ABZ-1607; **4, 5.** uralen UKU-1617-2u × canin UKU-1613; **6, 7.** uralen UKU-1618 × mutab ABZ-1665; **8.** uralen UKU-1617-4u × viridi UKU-1618; **9.** uralen ABZ-1631 × viridi UKU-1618; **10, 11.** uralen UKU-1617-2 × canin UKU-1617-3c; **12.** *E. viridiglumis* UKU-1618; **13, 14.** viridi UKU-1618 × uralens ABZ-1634; **15-18.** viridi UKU-1618 × canin UKU-1613; **19.** *E. mutabilis* ABZ-1607; **20, 21.** mutab ABZ-1607 × uralens ABZ-1634; **22.** *E. caninus* UKU-1613; **23.** canin UKU-1613 × mutab ABZ-1665; **24.** canin ABZ-1654 × mutab ABZ-1665.

1. In the combination ***ural × mutab*** (Fig. 2, tracks 2, 3, 6, 7) 4 plants were obtained. With the use of M11 primer 2 of 4 plants had paternal ~ 2000 b.p. band, and 2 other plants had paternal ~ 700 b.p. band (shown by arrows). Based on these results, we can conclude, that all 4 obtained plants are hybridous.
2. In the combination ***ural × canin*** (Fig. 2, tracks 4, 5, 10, 11) also 4 plants were obtained. With HB12 and M11 primers all 4 plants had fragments, specific for paternal form *E. caninus*, ~ 900 b.p. and ~ 800 b.p., respectively (shown by arrows). Hence, all 4 plants are hybridous.
3. In the combination ***ural × viridi*** (Fig. 2, tracks 8, 9) 2 plants were obtained. With M2 primer, the first plant had paternal ~ 850 b.p. band, and with HB12 primer it had paternal ~ 200 b.p. band (shown by arrows). With M11 primer both plants did not differ from the maternal form. We can conclude that the first plant is a hybrid, and the second one – is non-hybridous.
4. In the combination ***viridi × ural*** (Fig. 2, tracks 13, 14) all 2 obtained plants with primers HB12, M2 and M11 had amplicons, specific for paternal form *E. uralensis*, ~ 1100 b.p., ~ 600 b.p. and ~ 1200 b.p., respectively, what verifies that they are hybridous.
5. In the combination ***viridi × canin*** (Fig. 2, tracks 15-18) all 4 obtained plants with primers HB12 and M11 had specific paternal bands ~ 900 b.p. and ~ 800 b.p., respectively. With M2 primer paternal band in all plants coincides in size with the maternal one, so plants did not differ from the maternal form, but we can surely say, that all 4 plants are hybridous.
6. In the combination ***mutab × ural*** (Fig. 2, tracks 20, 21) 2 plants were obtained. With M2 and M11 primers these plants do not differ from maternal forms and do not have specific paternal amplicons. With HB12 one plant had double ~ 1000 b.p. fragment as in the maternal form, while the other plant does not have it. Based on a molecular data, we can assume that these plants are non-hybridous.
7. In the combination ***canin × mutab*** (Fig. 2, tracks 23, 24) all 2 obtained plants with primers HB12 and M2 had specific paternal bands ~ 600 b.p. and ~ 850 b.p., respectively. With M11 primer these plants had no clear paternal bands sized 2000 b.p., but they had paternal bands ~ 700 b.p. It all points that these plants are hybridous.

We have obtained hybrid F_1 plants in crosses of *E. uralensis*, *E. viridiglumis*, *E. mutabilis* and *E. caninus* in different combinations, except for the pair *E. mutabilis* – *E. viridiglumis*. The presence in hybrids of species-specific ISSR-bands from both parental forms is provided by high level of homozygosity of species in the genus *Elymus*, as a result of predominant self-pollination. Hence, the present study shows the possibility to determine the hybridity of plants at early ontogenesis stages by means of ISSR-markers. Subsequently, based on the results of the analysis of anthers condition and seed fertility of these hybrids, the conclusion will be drawn about the presence or absence of reproductive isolation and introgressive processes between species.

The work was carried out within the framework of the State Assignment of CSBG SB RAS "Assessment of morphogenetic potential of North Asian plant populations by experimental methods" (№ AAAA-A17-117012610051-5) with partial financial support of Russian Foundation for Basic Research (grant no. 18-04-01030). The publication was prepared with material of CSBG representing USU (United Specific Unit) "Collections of living plants indoors and outdoors", USU № 440534.

References

1. D.R. Dewey, *Gene manipulation in plant improvement* (Springer, Boston, MA, 1984)
2. N.S. Probatova, V.Y. Barkalov, A.V. Agafonov, *Scholary notes of Transbaikal State University. Series Biological Sciences*, **12**, 88-95 (2017) (in Russian)
3. A.V. Agafonov, *Rus. J. Genetics*, **40**, 1229-1238 (2004)

New locations of rare lichen species in the Krasnoyarsk region

Natalya Eromasova[1,**], and *Olga* Zyryanova[1]

[1]Khakas state University n. a. N. F. Katanov, 655000, Abakan, Russia

Abstract. As a result of lichenological studies of the Salbinka River valley, new locations of lichen species listed in the Red Data Books of the Russian Federation and the Krasnoyarsk Territory have been identified. New habitats of *Usnea longissima* Ach., *Lobaria isidiophora* Yoshim., *Lobaria scrobiculata* (Scop.) P. Gaertn. and *Lobaria pulmonaria* (L.) Hoffm. have been revealed. Their brief characteristics in the studied habitats are given and the state of thalli is described.

1 Introduction

6 years have passed since the last edition of the Red Book of the Krasnoyarsk Territory was published. Since that time, data on the distribution of lichen species under protection have been added, which will undoubtedly be taken into account in the development of the next edition. Recent lichenology studies focused directly on rare species are concentrated in the Western Sayan [1-3]. It is worth noting that the majority of lichen species, for which the locations in the Ermakovsky area are marked, are confined to taiga forests [4]. This article includes data on the new locations of rare species included in the Red Data Books of the Russian Federation [5] and the Krasnoyarsk Territory [4], discovered during lichenogical studies [6-7] in the Salbinka River valley (Ermakovskiy district, Krasnoyarsk Territory).

2 Materials and Methods

The studies were conducted in the spring-autumn period from 2015 to 2017 using the route method of floristic studies [8]. The territory of our study is located in the subtaiga subzone, where the dominants in the tree layer are *Betula pendula* Roth. and *Pinus sylvestris* L., in some places with an admixture of *Populus tremula* L. The species of lichens mentioned above, according to the known literary data, were not marked in such phytocenoses. The nomenclature of lichens is listed in accordance with the "List of lichen flora of Russia" [8], as well as the sites of the Consortium of North American Lichen Herbaria and Index Fungorum.
In determining the species, a binocular microscope MBS-10, microscope Micromed-1, reagents: 10% KOH solution, solution I (J + KJ) and paraphenylenediamine (C6H4 (NH2)

* Corresponding author: eromasova@inbox.ru

2) were used. The collections are stored in the Scientific Herbarium of the KSU. N.F. Katanova (HGU).

3 Results and discussion

In the study area, 4 species of lichens listed in the Red Data Books of the Russian Federation and the Krasnoyarsk Territory – *Usnea longissima* Ach., *Lobaria isidiophora* Yoshim., *Lobaria scrobiculata* (Scop.) P. Gaertn., *Lobaria pulmonaria* (L.) Hoffm. were found.

All of them, except *Lobaria pulmonaria* (L.) Hoffm., were found in new habitats, not typical for their usual location.

Usnea longissima Ach. It grows in humid areas of the region, mostly in dark coniferous taiga and mountain taiga old-growth frontier forests. It settles on the branches of coniferous trees. Common in the Sayans [4]. In the investigated territory, this species was found in a birch forest, on *Betula pendula* Roth., although it is not typical for it, as it is found only in coniferous trees in taiga forest conditions. Thallus, due to the natural and climatic conditions of the study area, differs from taiga specimens in smaller sizes, fibrils are also shorter.

Lobaria isidiophora Yoshim. This species is confined to dark coniferous mountain taiga old-growth forests, found on the Kulumys ridge and in the vicinity of the village of Tanzibey [4]. In the study area the species is marked in birch-pine and birch forests. It occurs mainly on old, often already dead trunks of species of the genus *Salix* L. Lichen is distributed unevenly in the territory. The largest concentrations are observed in places with high humidity: near the Salbinka River, streams and wetlands. Young thalli are rare, most of the detected specimens are old lichens with damage of varying degrees.

Lobaria scrobiculata (Scop.) P. Gaertn. It is confined to dark coniferous mountain-taiga forests of the Sayans, less often to subtaiga. In the Ermakovsky region it is noted for mountain-forest territories of the Sayans [4]. On the study area is found in birch forests, rarely. Most often these are small solitary thalli, which are adjacent to specimens of other species of Lobaria. Damage to the thallus was not noticed. As a forrophyte, *Betula pendula* Roth., Was also found on a moss covered boulder. Damage to the thalli was not noticed. As a forrophite, it uses Betula pendula Roth., also found on a moss covered boulder.

Considering the above species, *Lobaria pulmonaria* (L.) Hoffm should also be noted, with no new locations for it. It is distributed throughout the south of the Krasnoyarsk Territory to the latitude of Krasnoyarsk: in subtaiga, taiga and dark coniferous mountain taiga, Minusinsk depression [4]. Among the protected species, this is the most abundant and evenly distributed species in the study area. It prefers *Betula pendula* Roth. as forophyte, it can also be found on species of the genus *Salix* L., was singularly found on *Populus tremula* L. It was marked that this lichen grows mainly on trees, the trunk of which has a strong slope relative to the soil level. We believe that this regularity is associated with a longer retention of moisture by the trunks that are closer to the horizontal position. In general, individuals grow in small groups, rarely singly or form a continuous cover from the base of the trunk to 1.5 m in height on old birches. Completely undamaged thalli are rare, most are in a collapsing state, and this applies not only to old thalli, but also to very young ones.

4 Conclusions

Thus, we investigated the new locations of four protected species: *Usnea longissima* Ach. and *Lobaria scrobiculata* (Scop.) P. Gaertn. are found in birch forests; *Lobaria isidiophora*

Yoshim. – in birch and birch-pine forests; *Lobaria pulmonaria* (L.) Hoffm. on the investigated territory is marked in birch and birch-pine forests. Most of the thalli of *Lobaria isidiophora* Yoshim. and *Lobaria pulmonaria* (L.) Hoffm. are in a collapsing state. To this state of thalli of the species examined, perhaps, the increasing anthropogenic factor led, in the form of a close location of the federal highway P257 "Yenisei" and populated areas. Also in the study area and nearby logging is conducted.

We are grateful to Ph.D in Biology Kharitonenkov Maxim for assistance in preparing the publication.

References

1. T.N.Otnyukova, N.V. Stepanov, Modern Mycology in Russia **2**, 532-533 (2008)
2. N.V. Stepanov, The Bulletin of KrasGAU **4**, 47-52 (2010)
3. N.V. Stepanov, The Bulletin of KrasGAU **3**, 163-170 (2017)
4. *Red Book of the Krasnoyarsk territory. The Rare and Endangered Species of Wild Plants and Funguses* (Krasnoyarsk, 2012)
5. *Red Book of the Russian Federation (Plants and Funguses).* (Moscow, 2008)
6. N.I. Eromasova, Ecology of the Southern Siberia and adjacent territories **20**, 21-22 (2016)
7. N.I. Eromasova, Juvenis scientia. **4**, 12-14 (2017)
8. *Lichen flora of Russia: Biology, ecology, diversity, distribution and methods '3 study of lichens.* (Moscow; Saint-Petersburg: Association of scientific publications KMK, 2014)

Effect of methyl jasmonate on production of 20-hydroxyecdysone and turkesterone in *hairy roots* of *Silene linicola* C.C.Gmelin

Anna A. Erst [1,*], *Larisa* N. Zibareva[2], and *Elena* S. Filonenko[2]

[1]Central Siberian botanical garden of SB RAS Zolotodolinskaya st., 101, Novosibirsk, 630090, Russia
[2] National research Tomsk state university, Lenin avenue, 36, Tomsk, 634050, Russia

Abstract. Methyl jasmonate (MeJ) affects the regulation of secondary metabolism, and it is considered to be a promising elicitor in the culture of cells, tissues and plant organs. High-performance liquid chromatography method was used to identify the composition of ecdysteroids in *hairy roots* of *Silene linicola.* MeJ was found to have a stimulating effect on ecdysteroid biosynthesis in this culture. Addition of MeJ at a concentration of 100 μM increased the biosynthesis of 20-hydroxyecdysone by 74% (day 3), and that of turkesterone by 35% (day 6). The share of turkesterone in total ecdysteroid content in the investigated samples was up to 60%, and the content of 20-hydroxyecdysone was up to 30%. The study shows that MeJ is a promising stimulator of ecdysteroid biosynthesis in *hairy roots* of *S. linicola.*

1 Introduction

In recent years, a great deal of attention has been paid to the family Caryophyllaceae Juss. as a new source of ecdysteroids due to the discovery of this class of compounds in its composition. The representatives of this family are characterized by diverse composition of ecdysteroids, the presence of many new compounds not found in any other species, and high content of major components of total ecdysteroids [1, 2]. These species are promising biotechnological objects. The most common ecdysteroids in Caryophyllaceae Juss. are 20-hydroxyecdysone (20E), polypodine B, 2-deoxy-20-hydroxyecdysone, ecdysone, viticosterone E, and integristerone A.

A number of studies aim to develop new methods for increasing the biosynthetic ability of *in vitro* cultures. Some scientists suggest not to activate one or more key genes of secondary metabolism, but to develop approaches to remove inhibitor blocks and to shift the regulatory balance towards activation of the entire biosynthetic pathways [3]. Other scientists investigate the processes of enhanced secondary metabolism of *in vitro* cultures using proteomic methods that enable regulation of enzymes involved in biosynthesis of secondary metabolites [4]. However, most of the studies on regulation of *in vitro* bioactive compound accumulation focus on optimization of media and culture conditions, and on

* Corresponding author: annaerst@yandex.ru

selection of growth regulators, elicitors and precursors of secondary metabolite synthesis [5, 6]. Methyl jasmonate (MeJ) is one of the elicitors widely used in the culture of cells, tissues and plant organs [7–10]. It is shown that it has a stimulating effect on the content of ecdysteroids, in particular 20E, in plants *in vitro* and *ex vitro* [11, 12].

The current study aims to develop a strategy for elicitor-stimulated biosynthesis of phytoecdysteroids in *hairy roots* of *Silene linicola*, in particular 20E and turkesterone.

2 Materials and methods

2.1 Elicitation of *hairy root* cultures

Hairy roots of *S. linicola* were cultured on liquid B_5 medium [13] with the addition of 500 mg·l^{-1} casein hydrolysate at 30–42 day interval [14, 15]. Elicitation of ecdysteroid accumulation was carried out by adding MeJ to media at the 9th passage at the end of subculturing (day 30) at a concentration of 100 or 200 μM (22.43 or 44.86 mg of MeJ was dissolved in 4 ml of 70% ethanol, respectively) together with 30 ml of B_5 medium (ethanol concentration of 0.28%). To assess the effect of MeJ on growth characteristics of culturing and ecdysteroid biosynthesis, some of the samples were cultured only with the addition of ethanol solvent (at a concentration similar to that in the experiment with MeJ) and 30 ml of B5 medium. The samples were analyzed after 3, 6 and 12 days, after the components were added to media (on culture day 33, 36 and 42, respectively). Two controls were used in the study. Control 1 (C1) was the sample of hairy roots at the beginning of the experiment (culture day 30), control 2 (C2) was the sample at the end of the experiment (culture day 42), with no MeJ added.

2.2 Methods of research of biologically active substances of *hairy root* cultures

Extracts from 1–2 g of *hairy root* cultures were produced through repercolation with 70% ethanol solution followed by concentration to 2 ml with IKA rotary evaporator HB 10 digital (Germany). The analysis of phytoecdysteroids was performed in the Laboratory of Phytochemistry, SBG TSU (Tomsk, Russia) using high performance liquid chromatography according to the procedure described in [14, 15].

3 Results

3.1 Identification of ecdysteroids in ethanol extracts of *hairy root* cultures by HPLC

Turkesterone, 20E and polypodin B were identified in ethanol extracts of *hairy root* cultures by comparing the retention times of standards and peaks of compounds with absorption maxima characteristic of ecdysteroids at λ 240–250 nm. 20E was found in all the samples, turkesterone was found in virtually all samples except for sample 200-12, whereas polypodin B was detected only in control sample C2 on culture day 42. HPLC analysis of the investigated samples of *hairy roots* of *S. linicola* showed that a number of ecdysteroids, including nonpolar ones, are synthesized in the retention time interval of 23.7–51.0 min under the preditermined experimental conditions.

3.2 Quantitative determination of ecdysteroids in extracts of *hairy root* cultures

The results of the quantitative analysis of the ecdysteroid content in the samples in terms of the equivalent amount of absolutely dry raw materials are provided in the table.

Table. Ecdysteroid content in extracts of *hairy roots* of *S. linicola*.

No.	Samples	20E content, % (x ± m)	Turkesterone content, % (x ± m)	Total ecdysteroid content, %
1	C1	0.023±0.0008	0.016±0.0003	0.063 ± 0.0006
2	C2	0.014±0.0004	0.012±0.0005	0.055 ±0.0005
3	0-3	0.019±0.0002	0.045±0.0013	0.100 ± 0.0008
4	0-6	0.018±0.0005	0.054±0.0009	0.094 ± 0.0007
5	0-12	0.016±0.0003	0.028±0.0012	0.073 ± 0.0008
6	100-3	0.033±0.0009	0.038±0.0037	0.107 ±0.0023
7	100-6	0.030±0.0011	0.073±0.0023	0.138 ± 0.0017
8	100-12	0.011±0.0003	0.046±0.0007	0.134± 0.0005
9	200-3	0.025±0.0012	0.010±0.0003	0.056 ± 0.0008
10	200-6	0.018±0.0007	0.023±0.0014	0.061 ± 0.0011
11	200-12	0.021±0.0003	–	0.038 ±0.0003

Note: C1 – control, start of the experiment; C2 – control, end of the experiment (day 12); 0-3 – ethanol, day 3; 0-6 – ethanol, day 6; 0-12 – ethanol, day 12; 100-3 – 100 μM MeJ, day 3; 100-6 – 100 μM MeJ, day 6; 100-12 – 100 μM MeJ, day 12; 200-3 – 200 μM MeJ, day 3; 200-6 – 200 μM MeJ, day 6; 200-12 – 200 μM MeJ, day 12.

As can be seen from the table, the content of 20E in the control decreased from 0.023 to 0.014%. The content of 20E in the samples with no elicitor, but with ethanol added, was observed to decrease as well. However, this led to the active synthesis of turkesterone, which attained its maximum on day 6.

The addition of 100 μM MeJ results in an increased content of 20E on day 3 as compared to that in the control and samples cultured with ethanol. The pattern of changes in turkesterone after the addition of 100 μM MeJ is similar to that in the samples with ethanol. The maximum concentration of ethanol and total ecdysteroid content (table) can be observed on culture day 6. However, the comparison of turkesterone concentrations showed that its content in sample 100-6 exceeds that in the control by a factor of 4.5–6, and in sample 0-6 – by a factor of 1.4. In control samples, turkesterone amounts to 25% of total ecdysteroid content, whereas in samples with ethanol and those with 100 μM MeJ, it attains 40–60% and 35–52%, respectively. At the same time, the content of 20E in total ecdysteroids varies from 8 to 30%.

Weak inhibition of ecdysteroid synthesis can be observed in the experiment with MeJ added at a concentration of 200 μM (table).

4 Discussion

MeJ is a widely used elicitor in the culture of cells, tissues and plant organs *in vitro*. Thus, the addition of MeJ caused a 2-3-fold increase in the content of 20E in suspension and *hairy root* cultures of *Ajuga turkestanica* (Regel) Briq. [11]. A positive effect of MeJ on the accumulation of phytoecdysteroids was also observed for intact plants of *Spinacia oleracea*

L. [12]. The results of our studies revealed a stimulating effect of MeJ on ecdysteroid biosynthesis in *hairy roots* of *S. linicola*. It should be noted that the use of sodium acetate (the precursor of ecdysteroid biosynthesis) resulted in a different composition of ecdysteroids in *hairy roots* of *S. linicola* [14]; in particular, a valuable and rare turkesterone was not identified. However, ecdysone and 2-deoxy-20-hydroxyecdysone and 2-deoxy ecdysone typical of Caryophyllaceae Juss. were found in *S. linicola*.

The ecdysteroid composition in intact plants of *S. linicola* was found to include 20E, polypodine B, ecdysone, 2-deoxy-20-hydroxyecdysone, viticosterone E, turkesterone, and integristerone A [16]. In the introduced species of *S. linicola*, the major component is 20E, and the 20E to turkesterone ratio is 367:1 [16]. The study has shown that when 100 μM MeJ is supplemented, the content of turkesterone in *hairy roots* of *S. linicola* exceeds that of 20E by a factor of 1.7–3 on day 6. Hydroxylation of 20E is likely to occur to the 11C position, which leads to turkestherone synthesis, or to the 5C position, which causes polypodine B synthesis, and other positions of the steroid nucleus. At the same time, the content of the initial component (20E) decreases, but the content of other components is observed to increase.

Thus, *hairy roots* of *S. linicola* can be the source of a rare ecdysteroid, turkesterone, in the experiment scaling. To date, it has been identified in 5 plant species only: *A. turkestanic*, *Vitex glabrata*, *Vitex canescens* (Lamiaceae), *Leuzea carthamoides* (Asteraceae), *S. linicola* (Caryophyllaceae), and *Tapinella panuoides* (Fungi) (http: ecdybase.org/). Addition of MeJ at a concentration of 100 μM is a promising technique for stimulating biosynthesis of such valuable ecdysteroids as 20E and turkesterone in *hairy roots* of *S. linicola*.

The work was carried out with financial support of the budgetary project of Central Siberian Botanical Garden, SB RAS Assessment of morphogenetic potential of North Asian plant populations by experimental methods (AAAA-A17-117012610051-5). We also acknowledge the support of the RFBR grant, project № 16-44-7000634 Study of medicinal plants of Tomsk region and adaptive capabilities of introducents in order to create a raw material base of species, which are promising for treatment of socially significant diseases.

The study employed in vitro material from collection No. USU_440534 Collection of living plants indoors and outdoors.

References

1. L. Zibareva, V. Volodin, Z. Saatov, T. Savchenko, P. Whiting, R. Lafont, L. Dinan, Phytochemistry **64**, 499-517 (2003)
2. L. Zibareva, O. Volkova, S. Morozov, E. Chernyak, Phytochemistry **1**, 71-75 (2017)
3. V. Bulgakov, T. Avramenko, Biotechnol Lett. **37**, 1719-1727 (2015)
4. M. J. Martínez-Esteso, A. Martínez-Márquez, S. Sellés-Marchart, J. A. Morante-Carriel, R. Bru-Martínez, Front Plant Sci. **6**, 504 (2015)
5. A. Giri, M. L. Narasu, Biotechnology Advances **18**, 1-22 (2000)
6. M. I. Georgiev, J. Weber, A. Maciuk, Appl Microbiol Biotechnol **83**, 809-823 (2009)
7. L. Shabania, A. A. Ehsanpoura, G. Asgharib, J. Emamib, Russian Journal of Plant Physiology **56**, 621-626 (2009)
8. K. P. Martin, A. Sabovljevic, J. Madassery , J. Crop Sci. Biotech. **14**, 205-212 (2011)
9. M. Ram, K. V. Prasad, S. K. Singh, B. S. Hada, S. Kumar, Plant Cell Tiss. Organ Cult., **113**, 459–467 (2013)

10. B. Chodisetti, K. Rao, S. Gandi, A. Giri, In Vitro Cell.Dev.Biol.—Plant **51**, 88-92 (2015)
11. D. M.Cheng, G. G.Yousef, M. H.Grace, R. B.Rogers, J. Gorelick-Feldman, I. Raskin, M. A. Lila, Plant Cell Tiss. Organ. Cult. **93**, 73-83 (2008)
12. E. A. Schmelz, R. J.Grebenok, T. E.Ohnmeiss, W. S. Bowers, Chem. Ecol. **26**, 2883-2896 (2000)
13. O. L. Gamborg, D. E. Eveleigh, Can J Biochem. **46**, 417-421 (1968)
14. A. A. Erst, T. V. Zheleznichenko, A. A. Badulina, L. N. Zibareva, O. V. Kovzunova, Plant Cell Biotechnology and Molecular Biology **17**, 326-334 (2016)
15. A. Erst, L. Zibareva, T. Zheleznichenko, O. Kovzunova, Tomsk State University Journal, Biology **37**, 17-30 (2017)
16. L. N. Zibareva, *Phytoecdysteroids of plants of the family Caryophyllaceae* (Germany – LAP Lambert Academic Publishing GmbH&Co, Saarbrücken, 2012)

Application of phytoncidal plants for improvement of ecological parameters of the air of library interiors

Tatiana Fershalova[1,2*], *Elena* Baikova[1], and *Alina* Ushakova[2]

[1]Central Siberian Botanical Garden of the Siberian Branch of the Russian Academy of Sciences (CSBG SB RAS), Novosibirsk 630090, Russia
[2]Siberian state University of Geosystems and Technologies (SSUGT), Novosibirsk, 630108, Russia

Abstract. The results of the experiments on optimization of library room air with the help of the biological method are presented. Living tropical plants, volatile biologically active substances of which possess distinct antimicrobic properties are used for air sanitation. Their influence on change of the qualitative and quantitative microorganism composition in the air is shown. After plant placement the total number of colonies forming units (CFU) decreases by 3 times, that of staphylococcus – by 4.5 times. Air humidity increases insignificantly reaching the low limit of the comfort range for a person.

1 Introduction

In the world practice there is a wide experience of study of the ecological parameters of rooms of various functional purposes. The main ecological factors are temperature, humidity, air movement, qualitative and quantitative air microflora composition. These indices depend on many parameters: locality of a building, quality of used construction and finishing materials [1], as well as stay of people in the building [2]. People often spend 80% - 90% of time indoors [4]. In the indoor air environment there are many microorganisms – bacteria, house fungi and viruses [3]. The person passes through himself by breath on the average 14 m^3 of air per day [5]. He absorbs and secrets therewith volatile organic compounds (VOC), both endogenous and exogenous. Air quality indoors affects human performance and health as a whole [2,6,7,8]. Children are especially exposed to the risk of infection caused by poor air quality indoors [9,10]. They are more vulnerable to all types of environmental pollution compared to adults, as they breathe more air relative to body weight, the respiratory tract of children is not sufficiently developed [2,11]. Consequently, careful control of air and prevention of air pollution are essential in the premises of children's institutions [12]. Poor-quality indoor air of the libraries creates conditions for the appearance at employees and visitors of such diseases as: allergy, rhinitis, asthma and conjunctivitis [13]. Thus, a constant monitoring of the environmental factors promoting growth of air microorganisms and timely elimination of them are needed. There are various

* Corresponding author: fershalova@ngs.ru

ways of cleaning indoor air. Usually recommendations on the improvement of air are reduced to airing of rooms and sanitary processing. The aim of our work is optimization of the air environment of library interiors with the help of ecological phytodesign method developed on the base of Central Siberian Botanical Garden (CSBG SB RAS) [14]. This method is ecologically safe, extremely economical and aesthetic. The essence is to use plants with distinct antimicrobic and gas-absorbing properties to reduce microbic and chemical air pollution.

2 Material and methods

The experiment was conducted in the building of the city library where educational activities with participation of children were carried out regularly. Two rooms similar in volume ($195m^{3)}$, air temperature (+ 25 C^0), air humidity (20%) and illumination (500-2000 lx) were chosen. The windows faced the north-west. The experiment was conducted in the first room – hall of children's literature, the second room – literature issue hall was control. Air samples were taken in triplicate in five points for three days. The total microbic number of CFU, including *Staphylococcus* and *moldy mushrooms* in 1 m^3 of air (CFU/m^3) was counted. When making experiments, the air samples were taken before classes (20 children at 10-12 years old, duration of a class – 30 min) and after them in both halls - with plants (experiment) and without them (control). Air intake was performed by the aspiration method based on the forced deposition of microorganisms on the surface of a dense nutrient medium. With the help of the sampling device (SD-1B) were pumped 100 l of air which fell on the surface of Petri dishes with meat-peptone agar (MPA). The samples were selected at the height of 1,5–1,8 m from the floor – at the level of human breathing. Cultivation of microorganisms was carried out in the thermostat at +37C^0 for 2 days. Microclimatic parameters - temperature, air humidity and illumination were measured by the combined device "TKA-PK" (UF). A phytomodule containing plants grown in the green houses of CSBG SB RAS and adapted to the interior conditions was set in the experimental room. The assortment of plants was selected according to techniques of ecological phytodesign [7]: *Begonia* «Loospe», *B.* «Loospe – tu», *B.* «Nigramarga», *B.* «November Frost», *B.* «Major», *B.* «Nigramarga», *B.* «Orange County», *B. carolinifolia* Reg., *B.* x *erytrophylla* Neumann, *B. heracleifolia* Cham. et Schlecht., *Nephrolepis exaltata* (L.) Schott «Fluffy Ruffles», *Rhoeo discolor* (L'Her.) Hance), *Ruellia caroliniensis* (J.F.Gmel.) Steud.), *Pelargonium peltatum* (L.) L'Her. ex Ait., *Spathiphyllum wallisii* Reg., *Ficus Binnendijkii* Miq «Alii», *F. retusa* L., *Chlorophytum amaniense* Engl. «Green Orange», Ch. *comosum* (Thunb.) Jacques «Bonnie», *Ch. comosum* «Variegatum», *Schefflera octophylla* (Lour.), *Epipremnum aureum* (Linden & André) G.S.Bunting, «N'Joy». Calculation of the area of leaf surface of the phytomodule plants was made by the program "Siams photolab", the results were treated by "Microsoft Office Excel".

3 Results

In the course of the experiment air samples were taken in the rooms and the analysis of microorganisms grown on the nutrient medium was performed. The area of the leaf surface of plants required for getting sanitation effect was calculated. The optimal leaf area was 1,5-3 m^2 per 100 m^3 of air [7]. The leaf area in plants of the phytomodule amounted to 6,2 m^2.

On the first day of the experiment, air composition in the control and experimental rooms was practically the same (Fig.1, sample I). Twenty four hours later CFU/m^3 in the room with the phytomodule decreased by three times (Fig.1, sample II). Forty eight hours later the difference in CFU/m^3 between the experiment and the control remained unchanged

(Fig.1 sample III). Presence of children for 30 minutes increased significantly the quantity of microorganisms in the room air (Fig. 1, control), however, placement of plants allowed to decrease essentially CFU/m^3 in a day.

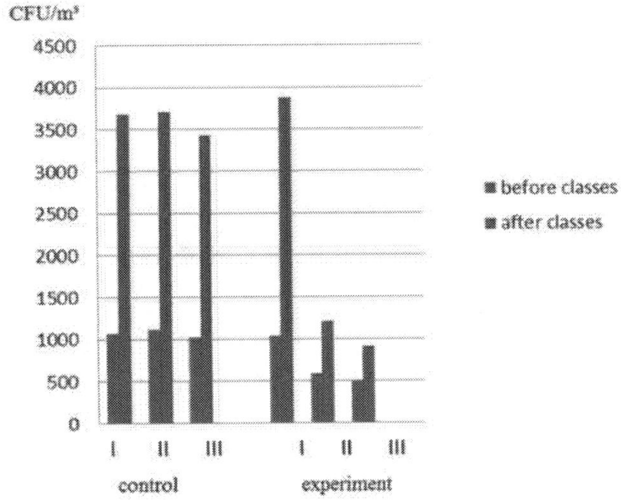

Fig. 1. Effect of plants on quantitative composition of air microflora in the room.

Figure 2 shows that the ratio of conditional-pathogenic microorganism microflora to constant one changed in the course of the experiment. In the control room *Staphylococcus* was more on 14% than in the room with plants.

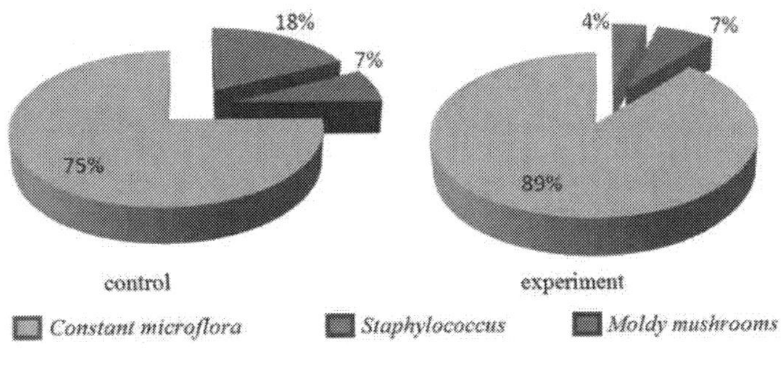

Fig. 2. Ratio of conditional-pathogenic microflora to constant one in the presence of plants (experiment) and without them (control).

It was revealed that air humidity in the library was lowered. Relative air humidity in the hall of children's literature before installation of the phytomodule was 21,6%. Forty eight hours after placement of plants air humidity amounted to 30,2% (Fig. 3).

4 Discussion

The ecological status of the room air environment is made up from quantitative and qualitative indicators: total number of microorganisms and ratio of conditional-pathogenic microflora to constant one. The microorganisms moving freely in the air belong to the constant microflora (*Micrococcus, Bacillus, Sarcina*) and the conditional-pathogenic one

(Staphylococcus, moldy mushrooms). Sanitary-hygienic state of a premise is characterized by the total number of all microorganisms in 1 m^3 of air - CFU/m^3.

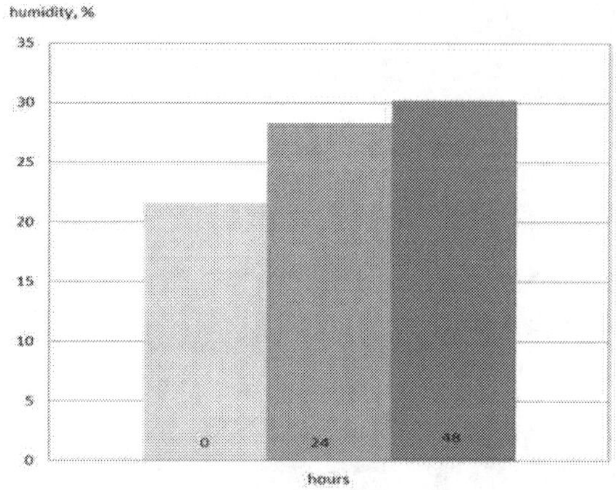

Fig. 3. Change in air humidity in the room after installation of the phytomodule.

In our experiment (control room) microorganism concentrations were in the range from 3435 to 3709 CFU/m^3 , which exceeded sanitary-hygienic standards (2000 CFU/m^3). But those indicators were not the most high. For instance, J. M. Daisey noted that in the course of experiments carried out in schools and kindergartens were found concentrations of vital bacteria in the range from 7 CFU/m^3 to 19500 CFU/m^3 [15]. Microorganism concentrations in the university library were significantly lower and varied from 367 to 2595 CFU/m^3 [5]. According to literature the number of bacteria in the room air environment increases due to presence of children [2]. We explain such an effect by the fact that adults are more static compared to children who often wave their hands, turn their heads, get up from their seats even during the classes. Such movements create chaotic flows of air currents in which there are different microorganisms. In the absence of air movement microorganisms gradually settle on the floor, and they are far less detectable at the level of breath of a person.

In the premises with library stock the main indicator of environmental well-being is quantity of *moldy mushrooms* colonies per 1m^3 of air as these microorganisms influence safety of books. The condition of air is considered satisfactory if quantity of *moldy mushrooms* per 1m^3 does not exceed 300–500 (CFU/m^3) [13, 16]. However, we have paid attention not only to quantity of *moldy mushrooms*, but also to presence of *Staphylococcus* in the air. These microorganisms do not influence safety of library collection, but are a source of various diseases of the person [17, 18, 19]. It was shown in our experiment that in the air of library rooms *Staphylococcus* was more than *moldy mushrooms*. Such data are confirmed by other authors. A. Fox, & R. M. T Rosario noted that concentration of gram-positive bacteria, to which *Staphylococcus* belongs, was usually higher than concentration of gram-negative bacteria [20, 21]. *Staphylococcus*, as a rule, gets to air from the mucosa of the respiratory tracts [18], but it may be on the surface of low-quality sanitary equipment [22] and even on the floor [21].

The results of our experiment show that application of ecological phytodesign method allows to improve air quality in premises of library. Thus, in a day after installation of phytomodules, value of CFU/ m^3 decreased significantly, air humidity reached 30.2% – the lower limit of the comfort range for a person (30% - 60%). However, such air humidity is still below the norm recommended for storage of books and documents – 45-55%. Each of premises has its own specifics, which should be taken into account when using the ecological phytodesign method. [14, 19, 22].

The work was carried out within the framework of the State Assignment of Central Siberian Botanical Garden SB RAS with partial support of Russian Foundation for Basic Research (project r_a №17-44-540601). When preparing the publication, the materials of bioresource scientific collection of CSBG SB RAS «Collections of living plants indoors and outdoors», № USU 44053 were used.

References

1. S. Torresin, G. Pernigotto, F. Cappelletti, et al., Indoor Air. Jul **28**,525-538 (2018)
2. N. Canha, S.M. Almeida, M.C. Freitas, et al., Arch. Environ. Prot. **41**, 11-22 (2015)
3. S.A. Wamedo, P.N. Ede, A. Chuku, Asian J. Biol. Sci. **5**, 183-191 (2012)
4. A.H. Awad, S.A. Farag, Int. J. Environ. Health Res. **9**, 313-319 (1999)
5. S.F. Hayleeyesus, A.M. Manaye, Asian Pac. J. Trop. Biomed. **4**, 312-317 (2014)
6. V.A. Ivanchenko, A.M. Grodzinsky, T.M. Cherevchenko, et al., *Phytoergonomics* (Naukova Dumka, Kiev, 1989)
7. N.V. Tsybulya, T.D. Fershalova, *Ecological basics of phytodesign* (SGGA, Novosibirsk, 2013)
8. D. P. Wyon, Indoor Air **14**, 92-101 (2004)
9. Yu.L. Yakimova, N.A. Rychkova, N.V. Tsybulya, Contem. Probl. Ecol. **2**, 249-253 (2002)
10. E. Bragoszewska, A. Mainka, J.S. Pastuszka, et al., Atmosphere **9**, 1-15 (2018)
11. P.T.B.S. Branco, R.A.O. Nunes, M.C.M Alvim-Ferraz, et al., Rev. Port. Pneumol. **22**, 293-295 (2016)
12. T. Soto, M. Rosa, G. Murcia, et al., Anales de Biología **31**, 109-115 (2009)
13. E.A. Agafonova, *Recommendations on the work with photo documents from the State funds of the Russian Federation* (State museum and exhibition center Russian Photo, St. Petersburg, 2012)
14. N.V. Tsybulya, T.D. Fershalova, *Phytoncidal plants in the interior. Air sanitation with the help of plants* (Novosibirsk Publishing House, Novosibirsk, 2000)
15. M. Daisey, W.J. Angell, M.G. Apte, Indoor Air **13**, 53-64 (2003)
16. K. Konsa, M. Siiner, Theory and practice of preservation of cultural monuments **17**, 9-16 (1995)
17. A.M. Andersson, N.Weiss, F. Rainey, et al., J. Appl. Microbiol. **86**, 622-634 (1999)
18. O.V. Beresneva, *New technologies in library and information practice and personneltraining* (Perm, 2006)
19. N.V.Tsybulya, Yu.L. Yakimova, N.A.Rychkova, et al., Rastitelnye resursy **38**, 112-117 (2002)
20. A. Fox, R.M.T. Rosario, Indoor Air **4,** 239-247 (1994)
21. U. Lignell, *Characterization of microorganisms in indoor environments* (Univ. Kuopio Fac. Nat. and Environ. Sci., Diss., 2008)
22. N.V. Tsybulya, T.D. Fershalova, L.P. Davidovich, Izv. Samara Sci. Center, RAS **19**, 360-364 (2017)

BIO Web of Conferences **11**, 00015 (2018) https://doi.org/10.1051/bioconf/20181100015
Prospects of Development and Challenges of Modern Botany

Fungal records database of Khanty-Mansi Autonomous Okrug – Yugra

Nina Filippova[1,*], and *Tatiana* Bulyonkova[2]

[1]Yugra State University, Environmental Dynamics and Global Climate Change-UNESCO chair, 628012, Chekhova street, 16, Khanty-Mansiysk, Russia
[2]A.P. Ershov Institute of Informatics Systems Russian Academy of Sciences, 630090, prospect Lavrentyeva, 6, Novosibirsk, Russia

Abstract. The mycological research in the Northern part of Western Siberia becomes quite developed by now to be reviewed, in Khanty-Mansi Autonomous Okrug – Yugra reaching about 150 scientific works accomplished mainly from the second half of XXI century. There is even greater need in accumulating fungal records reported from different publications in a standardized regional database. This database should be organized according to contemporary demands of biodiversity informatics, to be compatible and accessible by other researchers. Organized in 2016, the Fungal Records Database of Yugra, presently contains about 14000 records of 2500 species of fungi and fungi-like organisms recorded in the area. This paper describes the database organization and accumulated fungal records and introduces the research conducted in the field of mycology in Yugra from the beginning of the XX century up to present.

1 Introduction

The mycological research in Khanty-Mansi Autonomous Okrug – Yugra stems from isolated studies in the beginning of the XX century, yet regular and systematic research only began in the 70–80s. Over the following decades several dozens of researchers worked in the area, and a total of about 150 scientific works were published [1, 2]. However, these were scattered works which have not been summarized until now and there has not been a general checklist for the area. The recent development of biodiversity informatics opened a new way of accumulation species records in large standardized databases which can be considered as a substitute to printed checklists of older times. With this in mind, we decided to review the available publications, extract fungal records, and analyse the accumulated knowledge of fungal biodiversity in Yugra region according to the modern requirements to the biodiversity databases.

The goal of this paper is to describe the structure and the field format of the regional database of fungal records, make the analysis of accumulated presently data on fungal diversity in Yugra, and make an overview of the mycological research reflected in the database.

* Corresponding author: filippova.courlee.nina@gmail.com

2 Organization of a regional Fungal Records Database

The Fungal Records Database of Yugra (FReDY) is aimed at summarizing the species records accumulated in the course of previous mycological studies in Yugra. The database structure is based on Darwin Core format and includes about 30 fields describing taxa, record locations and other auxiliary information (table 1). Google Sheets, a Web-based service, was used as the basis for a simple database. The elements of the relational structure were created using reference tables and queries [3]. Such database can be filled simultaneously by many specialists, and a common data format will be provided.

Table 1. The structure of the database is represented by the following tables and fields.

Tables	Fields
Main Data	*recordNumber* – individual record number (increment numbers, formatted «FReDY-xxxxx»); *originalNameUsage* – the names of the species as reported in the source; *scientificName* – name of species, in accordance with modern taxonomy and nomenclature data (filled automatically with a query from the Species table); *identificationQualifier* – the quality of the identification (for notes like cf., aff.); *taxonRemarks* – notes about the discrepancy in spelling of the species names; *taxonRank* – the taxonomic rank of the entry given in the *scientificName* field; *genus* – genus (filled automatically with a query from the *scientificName* field); *group* – the traditional group (filled automatically with a query from the Taxonomy table, the term is absent in DwC); *family* – family (filled automatically with a query from the Taxonomy table); *order* – order (filled automatically with a query from the Taxonomy table); *bibliographicCitation* – reference to the literature source reporting the registration of the species; *stateProvince* – the highest administrative subdivision of Russia (language – English, source – Getty Thesaurus of Geographic Names, http://www.getty.edu/research/tools/vocabularies/tgn); *county* – rayon (language – English, source of names – Yandex.Maps); *locality* – location of the find (language – English, source of names – Yandex.Maps); *verbatimLocality* – the location of the find in the original (Russian) language as reported in the source; *protectedArea* – the territory of protected areas (term is absent in DwC); *habitat* – habitat (description of vegetation type, landscape and / or micro-conditions, language – English); *verbatimHabitat* – habitat in the original (Russian) language (term is absent in DwC); *catalogNumber* – herbarium number of the record, if reported in publication; *eventDate* – the date of registration of the find, if reported in publication; *decimalLatitude* – latitude (Decimal degrees); *decimalLongitude* – longitude (Decimal degrees); *geodeticDatum* – datum (most often WGS84); *coordinateUncertaintyInMeters* – the approximation of the coordinate (see DwC definition)
Synonymy	A list of species names and corresponding current synonyms (filled in manually by the community of specialists [3])
References	Bibliographic records (about 20 standard bibliographic fields)

Geography	Geographic tree from Rayon to Country (built on the basis of Getty Thesaurus of Geographic Names and Yandex.Maps)
Taxonomy	Taxonomic tree including group, order, family and genus (compiled manually based on a number of taxonomic sources)
Account	Used to obtain statistical information using queries, for example, unique lists of species for individual territories and taxonomic analyses

3 Analysis

Presently, the database contains over 14 000 fungal records in the region and adjacent areas, reported in 72 scientific publications. According to the database summary report, there are currently about 2500 species of fungi and fungi-like organisms recorded in Yugra. The richest groups are the agaricoid basidiomycetes (31%), lichens (33%) and aphyllophoroid basidiomycetes (19%). The less studied groups are the ascomycetes (8%), myxomycetes (3%), heterobasidiomycetous fungi (<1%), and yeasts <1%).

Regarding the territorial coverage of the records, all administrative divisions are represented in the database, however, the majority (72%) of collections were registered in three administrative divisions (Khanty-Mansiyskiy – 30%, Berezovskiy – 24% and Surgutskiy district – 18%) (figure 1).

The FReDY database was uploaded to GBIF (YSU BC publisher) by 11 November 2017; it is available for review of metadata and provides open access to the data upon download [4].

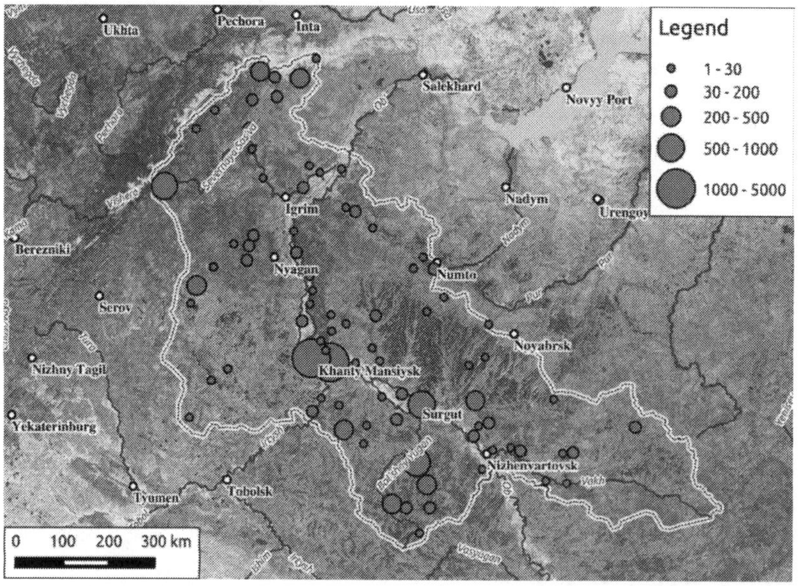

Fig. 1. The map of records in FReDY in 2017. The clustering of points was made within a radius of 20 km; the scale breaks were selected manually after plotting the frequency distribution histogram.

4 Overview of the mycological research reflected in the database

4.1 Lignicolous basidiomycetes

Lignicolous basidiomycetes are a well-studied ecological group in the region and about 35% of records in FReDY represented by this group. Viktor A. Mukhin analysed the geography and ecology of lignicolous basidiomycetes throughout all zones of Western Siberia from the forest-steppe zone in the South to tundra-steppe in the North. He theorized on the geographic aspect of the possible origin of the local mycota, outlined species distribution ranges and analysed ecological factors affecting them (e.g. humidity, temperature, substrate specialization, or interspecific competition). Communities of lignicolous fungi were analysed based on forest types: dark coniferous, light coniferous, broad-leaved, small-leaved or floodplain forest communities. The stages of wood decomposition in Western Siberia are described and characterized by the sets of species of the corresponding fungal community. Finally, rare species, for the whole territory or its particular habitats and zones, were revealed, complete with conservation recommendations.

Stanislav P. Arefyev and Iraida V. Stavishenko continued the study of lignicolous basidiomycetes in the region. They have contributed greatly to the knowledge of species diversity in most of the region's conservation areas and performed ecological monitoring in the areas of oil and gas extraction. I. V. Stavishenko described in detail the lignicolous macrofungi of the middle-taiga Ob basin region in her PhD thesis. S. P. Arefyev applied mathematical methods of ecological modelling to predict species composition of lignicolous fungi depending on substrate characteristics, climate and disturbance level of the forest.

4.2 Terricolous macrofungi

Terricolous macrofungi are a key ecological group of saprotrophs and ectomycorrhizal species and about 30% of records in FReDY represented by this group. The diversity of macrofungi was studied in a number of districts of KhMAO: the most thoroughly studied area is centred around the regional capital, Khanty-Mansiysk, and in the south-east part, in and near the Yuganskiy nature reserve). Nina V. Filippova with co-authors has been studying the species diversity and fungal community structure in forests near Khanty-Mansiysk while Elena A. Zvyagina and Tatiana M. Bulyonkova continuously monitor the fungal biodiversity in the Yuganskiy nature reserve. Anton G. Shiryaev has published a study of clavarioid basidiomycetes in a conservation area. In addition to the annotated species lists, a number of works deal with specific aspects of ecology, biology and abundance of rare, protected and otherwise noteworthy species.

4.3 Phytopathological studies

A variety of studies of plant pathogens, including pathogens of forests, have been carried out in the region. Stanislav P. Arefyev studied fungal pathogens of the Siberian pine (*Pinus sibirica*) and described the ecological causes of the diseases along with recommendations on the afforestation of Siberian pine cultures. Tatiana A. Makarova with co-authors made studies on the causes and extent of mycopathological diseases of urban spaces around the city of Surgut and in some other parts of the region.

4.4 Other groups of fungi and fungus-like organisms

The lichens represent a large part of fungal diversity described for the region and 20% of records in FReDY. The lichen mycota was described in several conservation areas, with the highest diversity revealed in the North Ural mountains, reaching around 900 species within a relatively small area (Nelly V. Sedelnikova).

The species diversity of myxomycetes was covered by two studies in conservation areas (Konstantin A. Fefelov).

The whole territory of the Khanty-Mansi Autonomous Okrug is strongly bogged, with peatlands covering up to 70% of the area. The fungal communities of peatlands have been studied in many aspects, including the diversity of macrofungi, microfungi on different litter debris of bog plants, yeasts on Sphagnum and bog plants, lignicolous fungi of bog-dwelling trees (the Scots pine) (Nina F. Filippova). Totally fungal records collected during the research in peatlands represent about 8% of records accumulated in the database.

The authors are grateful to Ilya Filippov for preparation of a graph.

References

1. N.V. Filippova, S.P. Arefyev, T.M. Bulyonkova, E.A. Zvyagina, V.I. Kapitonov, T.A. Makarova, V.A. Mukhin, I.V. Stavishenko, E.I. Tavshanzhi, A.G. Shiryaev, Environmental dynamics and global climate change **8**, 18-28 (2017)
2. N.V. Filippova, S.P. Arefyev, T.M. Bulyonkova, E.A. Zvyagina, V.I. Kapitonov, T.A. Makarova, V.A. Mukhin, I.V. Stavishenko, E.I. Tavshanzhi, A.G. Shiryaev, Environmental dynamics and global climate change **8**, 29-45 (2017)
3. S.Yu. Bolshakov, N.V. Filippova, K.O. Potapov, D.V. Ageev, S.V. Volobuev, *International Conference «The use of modern information technologies in botanical investigations»* (Apatity, 2017)
4. N. Filippova, S. Arefyev, T. Bulyonkova, E. Zvyagina, V. Kapitonov, T. Makarova, V. Mukhin, I. Stavishenko, E. Tavshanzhi, A. Shiryaev, Fungal records database of Khanty-Mansi Autonomous Okrug – Yugra (Russia, West Siberia). Yugra State University Biological Collection (YSU BC). Occurrence Dataset https://doi.org/10.15468/hfje3l accessed via GBIF.org on 2017-11-19 (2017)

BIO Web of Conferences **11**, 00016 (2018) https://doi.org/10.1051/bioconf/20181100016
Prospects of Development and Challenges of Modern Botany

Vitality structure of the endemic *Oxytropis kungurensis* Knjasev (Fabaceae) coenopopulation on the eastern shore of Lake Aushkul (Republic of Bashkortostan)

Gulnaz Galikeeva, *Olga* Elizaryeva, *Albert* Muldashev, *Natalia* Tyutyunova, and *Natalia* Maslova*
Ufa Institute of Biology. Ufa Federal Research Centre, RAS, 69, prospekt Oktyabrya, Ufa, 450054, Russia

Abstract. Consideration is given to biometric indices and vitality structure of *Oxytropis kungurensis* Knjasev (Fabaceae), an endemic species of the South Urals and Central Cis-Ural Region, observed in 2015 and 2016 in the coenopopulation on the eastern shore of Lake Aushkul (Uchalinsky District, Republic of Bashkortostan, Russia). The paper reveals differences in plant biometrics and vitality according to the years of observation. Judging from 14 biometric indices out of 19 indices analyzed, middle-aged generative plants in 2015 surpass those in 2016. Indices that characterize the vitality structure are higher in 2015. The vitality type of the coenopopulation varies from prospering to depressive. Variations in biometric and vitality indices are associated with changing weather conditions during the vegetation period. In 2015, there were optimal conditions for plants to grow and develop.

1 Introduction

Assessment of the vitality state is part of comprehensive research on ecological and biological features of rare plant species. Information about the vitality status of each particular plant provides opportunities for revealing the vitality structure of coenopopulations (CPs) and their total assessment based on the proportion of plants with different vitality [1]. The results of such investigations demonstrate the stability and dynamics of a population in general and also detect the ecotope.

2 Materials and Methods

2.1 Research Subject and Aims

The subject of the research is *Oxytropis kungurensis* Knjasev (Fabaceae), an endemic species of the South Urals and Central Cis-Ural Region. The species is included into the

* Corresponding author: maslovanv-ib-ufa@mail.ru

Red Data Book of the Republic of Bashkortostan (group 2, i.e. decreasing in the population number) as well as into the Red Data Book of the Perm and Chelyabinsk Regions [2]. In Bashkortostan, the species is found in 15 sites in the Uchalinsky and Beloretsky Districts. This mesoxerophyte occupies rock outcrops and stony steppes composed of limestones and mafic rocks (basalts, etc.) as well as highland tundras (Mashak Ridge). All known CPs of the species are usually scanty, and their current status is unsatisfactory [2]. The study of such species with a small number of localities and individual plants is highly topical [3-4].

The aim of this work is to study biometric indices and the vitality structure of *O. kungurensis* CPs on the eastern shore of Lake Aushkul in the Uchalinsky District of Bashkortostan. The research was conducted in 2015 and 2016. The CP area is about 180 m^2. The slope of south-western exposition is about 20° steep. Chernozem-like soils are very stony and poorly developed or organogenic rubbly. Vegetation is of the stony steppe type with thin grasses (projective cover is 35–40 %). The dominant role is played by petrophytic species characteristic of Trans-Uralian and Siberian steppes. Grazing is not frequent. *O. kungurensis* plants grow primarily on steppe plots covered with moss (projective cover is 10–20 %).

2.2 Research Methods and Statistical Analysis

Biometric indices were studied in middle-aged generative plants. The number of samples analyzed was 22 plants in 2015 and 29 plants in 2016. A comparison of mean indicator values by the years of observation was performed using the Student's t-test [5]. The research on the vitality structure was carried out using two methods. According to the Zlobin's method [1], the determinant traits were selected out of 19 morphometric characteristics on the results of factor analysis. A two-dimensional approach was used to group the plants according to their height and number rosette-forming shoots. The plants were classified into three groups: *a* – high vitality level, *b* – average vitality level and *c* – low vitality level. The quality of the CP was assessed using the integral index $Q=(a+b)/2$ [1]. The vitality of the CP was also determined using the method of Ishbirdin et al. [6]. The vitality indices of the CP (IVC) was calculated using the weighted average method according to 19 morphometric traits, and the ratio $I_q=(a+b)/2c$ was determined. Data statistical processing [6] was carried out for software packages MS Excel 2003 and Statistica 6.0.

3 Results and Discussion

A comparison of mean values of the biometric indices performed using the Student's t-test according to the years of observation showed reliable differences on the basis of 14 indices (Table 1). In 2015 these parameters were in 1.2 to 2.4 times higher than in 2016 (1.5 times higher on the average). For indices number of vegetative shoots, leaf lamina width, number of leaflet pairs, lateral leaflet length and width reliable differences was not showed.

Interannual differences in the biometric indices of the *O. kungurensis* middle-aged generative plants were also recorded in the CP 1.5 km south-westwards of Lake Aushkul when comparing the data for 2014 and 2015 [7].

The vitality status of plants and their CP varies depending on the year of observation. The year 2015 saw a prospering CP represented by three dimensional groups of plants (Table 2). The dominant plants were those of group *b* (0.41) and group *a* (0.50). Here, the quality index was maximal (0.45). Indices IVC and I_Q had high values as well. In 2016, the CP contained dominant plants with the low vitality status (group *c*) – 0.66. The portions of plants were 0.31 for the intermediate-status group and only 0.03 for the high-status group, respectively. Such conditions were responsible for the formation of a depressive CP. The quality of the CP was 2.6 times lower as compared to 2015. The values of IVC and I_Q indices were also low. The values of the quality index (Q) and vitality index (IVC) are in agreement with each other.

Table 1. Biometric indices of *Oxytropis kungurensis* middle-aged generative plants in the coenopopulation on the eastern shore of Lake Aushkul

Indices	2015	2016	$t_{actual.}$
Plant height, cm	25.1±1.0	20.3±0.9	3.732
Rosette height, cm	14.5±0.5	10.3±0.5	5.686
Rosette diameter, cm	21.8±1.1	13.2±0.7	6.603
Caudex diameter, cm	5.8±0.4	4.0±0.2	4.279
Number of inflorescences, pcs	9.2±0.7	6.1±0.5	3.504
Number of flowers per inflorescence, pcs	15.1±0.5	12.6±0.5	3.535
Number of flowers per plant, pcs	138.2±12.2	77.4±7.4	4.268
Number shoots per rosette, pcs	7.2±1.0	4.8±0.4	2.382
Number of vegetative-generative shoots, pcs	6.0±0.6	3.8±0.3	3.278
Number of leaves per plant, pcs	31.0±3.8	12.8±0.9	4.654
Number of leaves per floriferous shoot, pcs	5.0±0.3	3.3±0.2	5.505
Petiole length, cm	6.3±0.3	4.6±0.3	4.192
Leaf lamina length, cm	7.9±0.4	5.8±0.3	4.231
Leaf length, cm	14.3±0.6	10.4±0.6	4.561

Variations in the vitality indices are associated with different meteorological indicators during the vegetal periods in 2015 and 2016, primarily with the total precipitation in May and June. Anthropogenic effects on the CP remained the same. Long-term phenological observations in the Ufa Botanical Garden suggest that blooming and mass vegetative propagation occur in May and June [8]. This is a crucial period in the plant growth and development. Analysis of meteorological data showed that in 2015 the vicinities of Lake Aushkul received 187.8 % of the long-term average precipitation in May and 110.2 % in June (Figure 1), whereas in 2016 these values did not exceed 52.2 % in May and 53.6 % in June [9-10]. In 2015, there were optimal conditions for plant growth and development. Over the vegetal period the total precipitation comprised 328.5 mm in 2015 (87.8 % of the norm) and 259.9 mm in 2016 (69.5 % of the norm). In 2015, average monthly air temperature in May and June exceeded the average monthly norm, which exerted a favourable effect on plant growth rates. In 2016, average monthly air temperature during the same period was consistent with the long-term average. In 2016, the vital status of the plants was also affected by weather conditions occurred in August and September 2015. August was cool (average monthly air temperature was by 2.5°C below the norm) and September was dry (precipitation was 38.1 % of the norm). This is precisely the period when the inflorescence formation takes place in this species. In 2016, such indices as the number of flowers per inflorescence and the number of inflorescences, flowers and vegetative-generative shoots per plant were lower.

Table 2. Vitality indices of the *Oxytropis kungurensis* coenopopulation on the eastern shore of Lake Aushkul

Year of observation	Proportion of plants according to vitality groups			Vitality indices of the CP			
				Q	I_Q	IVC	Vitality type
	c	*b*	*a*				
2015	0.09	0.41	0.50	0.45	5.00	1.19	prospering

2016	0.66	0.31	0.03	0.17	0.26	0.86	depressive

The influence of ecological factors on the vitality of plants and CPs is diagnosed and studied in many studies [11-14]. For example, we determined the dependence of the vitality of plants and populations of *Oxytropis baschkirensis* Knjasev from weather changes [15].

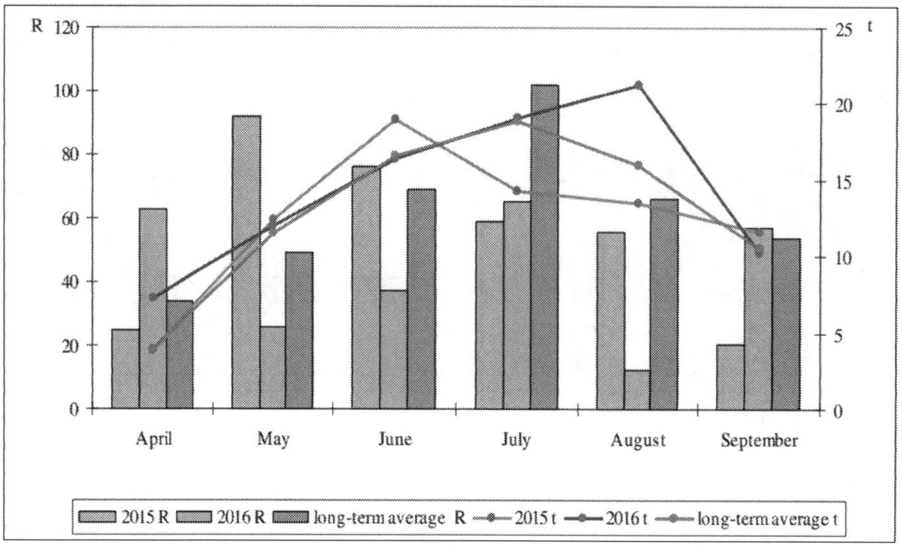

Fig. 1. Meteorological characteristics of the vegetation period in the area of investigation in 2015 and 2016. *X-axis* – months; *Y-axis*: *right* – air temperature, °C (t); *left* – precipitation, mm (R).

Thus, biometric indices of *O. kungurensis* tend to vary depending on the year of observation. These variations affect the proportion of plants with different vitality status and eventually change the vitality type of the population.

References

1. Yu.A. Zlobin, *Principles and methods study of plant coenotic populations* (Educ. Kazan Univ., Kazan, 1989)

2. *Red Data Book of the Republic of Bashkortostan*, **1** (MediaPrint, Ufa, 2011)

3. P.L. Gorchakovsky, V.N. Zueva, Rus. J. Ecol. **3** (1984)

4. T.J. Walisch, G. Colling, M. Bodenseh, D. Matthies, Ann. Botan. **115**, 7 (2015) doi:10.1093/aob/mcv040

5. G.N. Zaytsev, *Mathematical analysis of biological data* (Nauka, Moscow, 1991)

6. A.R. Ishbirdin, M.M. Ishmuratova, T.V. Zhirnova, Bull. N. Novgorod Univ. of N.I. Lobachevsky, Ser. Biol. **1**, 9 (2005)

7. G.M. Galikeeva, N.V. Maslova, Bull. Ufa SC RAS **4**, 1 (2015)

8. N.V. Maslova, O.A. Elizaryeva, G.M. Galikeeva, N.M. Tyutyunova, IOP Conf. Series: EES **107,** 012086 (2017) doi:10.1088/1755-1315/107/1/012086

9. http://www.pogodaiklimat.ru/ (2018)

10. https://ru.climate-data.org/location/1843/ (2018)

11. R. Graper, Rhodora **82**, 829 (1980)

12. O. Solbrig, Evolution **35**, 6 (1981)

13. M. Begon, M. Mortimer. *Population ecology. A unified study of animals and plants* (Blackwell Sci. Publ., Oxford, 1986)

14. J.D. Thompson., A.J. Gray, T. McNeilly, Acta Oecol. **11**, 5 (1990)

15. N.V. Maslova, A.A. Muldashev, O.A. Elizaryeva, A.Kh. Galeeva, D.N. Kuvatova, Modern Probl. of Sc. and Educ. **3** (2015)

BIO Web of Conferences **11**, 00017 (2018) https://doi.org/10.1051/bioconf/20181100017
Prospects of Development and Challenges of Modern Botany

Comparative analysis of chemical composition of *Miscanthus* var. Soranovskii

Yulia A. Gismatulina*, and *Vera V.* Budaeva

Institute for Problems of Chemical and Energetic Technologies, Siberian Branch of the Russian Academy of Sciences (IPCET SB RAS), 659322 Biysk, Russia

Abstract. The search for alternative sources of cellulose is currently getting special importance. *Miscanthus* holds promise as a cellulosic feedstock with a high potential for industrial-scale cultivation. The present study reports an analysis done for seven years of the chemical composition of the aboveground biomass of *Miscanthus* var. Soranovskii as a source for producing bioethanol. The aboveground biomass of *Miscanthus* was found to have a high cellulose content (about 50%), irrespective of the age, and it is therefore reasonable to utilize this cultivar for bioethanol production.

Today, of special relevance are efforts focused on rendering plant cellulosic biomass into bioethanol with subsequent production of ethylene and other valuable carbohydrate polymers are now gaining ground (Skiba, 2017). The choice of a raw material for the obtention of bioethanol is usually guided by the cumulative indicator—ethanol yield from 1 ha—which takes into account crop yielding capacity and productivity in ethanol depending on sugar, cellulose or starch content. In Russia, *Miscanthus* has high potential for breeding at an industrial scale. *Miscanthus* is a perennial, fast-growing energy crop that provides high biomass yields of 10-15 t/ha/year over a span of 15-25 years, with low requirements for growing conditions (Gismatulina, 2017). *Miscanthus* offers a range of advantages: high cellulose content (50%); unpretentiousness to soils and frost resistantance; low requirements for fertilizer application; and enhanced tolerance to diseases and pests. This complex of properties suggests a promising outlook for using *Miscanthus* to create a sustainable raw materials database for ethanol production.

Having regard to the above, this work was aimed at analyzing the chemical composition of the aboveground biomass of *Miscanthus* var. Soranovskii at each year over the span of seven years in order to assess the prospects of obtaining bioethanol therefrom.

The substrate for this study was seven *Miscanthus* var. Soranovskii (Slynko, 2013). To assess the prospects of producing bioethanol from *Miscanthus*, it was necessary to examine its chemical composition. For the examination, we took the aboveground biomass of *Miscanthus* var. Soranovskii over the span of seven years (2011-2017), which had been grown on experimental land plots of IPCET SB RAS (Biysk) completely without using agricultural methods (fertilization, watering, loosening and weeding).

The chemical compositions were measured by standard analytical procedures for plant biomass (Obolenskaya, 1991).

* Corresponding author: julja.gismatulina@rambler.ru

The work was performed using instruments of the Biysk Regional Center for Shared Use of Scientific Equipment of the SB RAS (IPCET SB RAS, Biysk).

Table 1 summarizes chemical compositions of the aboveground biomass of *Miscanthus* var. Soranovskii harvested at each year over the period of seven years.

Table 1. Chemical compositions of the aboveground biomass of *Miscanthus* var. Soranovskii for the period of 2011-2017

Harvest year /Plantation age	Content*, %				
	Kuschner cellulose	Ash	Lignin	Pentosans	Fat-wax
2011 / 1 year-old	42	6.30	22.23	25.33	5.71
2012 / 2 year-old	44	6.20	23.81	23.59	4.78
2013 / 3-year-old	45	4.62	21.11	25.10	2.81
2014 / 4 year-old	53	5.87	21.99	21.00	4.98
2015 / 5 year-old	54	3.57	20.13	18.57	3.57
2016 / 6 year-old	50	3.24	22.35	20.40	4.46
2017 / 7 year-old	50	5.06	23.08	20.44	4.86
Note: * – on an oven-dry basis.					

It follows from the tabulated data that *Miscanthus* var. Soranovskii holds promise as a cellulosic resource containing 42-54% Kuschner cellulose, therefore, the obtention of bioethanol therefrom is expedient. The Kuschner cellulose content was observed to rapidly increase over the initial five years of harvesting, from 42 % (harvest of the year 2011) to 54 % (harvest of the year 2015). The cellulose content then insignificantly declined from 54 % (2015 harvest) to 50 % (2016-2017 harvests) and remained at this level for two years. No regularities of the variation of the non-cellulosics content were noticed over the years of this study.

Overall, it was found in the course of the study on the chemical composition of the aboveground biomass of *Miscanthus* var. Soranovskii throughout seven years that *Miscanthus* has a Kuschner cellulose content of about 50%, and it is reasonable to utilize it for bioethanol production.

The research was supported by Integrated Program of Basic Research of the SB RAS II.1 'Interdisciplinary Integrated Research' (Project No. 0385-2018-0013).

References

1. E.A. Skiba, O.V. Baibakova, V.V. Budaeva, I.N. Pavlov, M.S. Vasilishin, E.I. Makarova, G.V. Sakovich, E.V. Ovchinnikova, S.P. Banzaraktsaeva, N.V. Vernikovskaya, V.A. Chumachenko, Chem. Eng. J. **329**, 178 (2017)
2. Yu.A. Gismatulina, V.V. Budaeva, Ind. Crop. Prod. **109**, 227 (2017)
3. N.M. Slynko, T.N. Goryachkovskaya, S.V. Shekhovtsov, S.V. Bannikova, N.V. Burmakina, K.V. Starostin, A.S. Rozanov, N.N. Nechiporenko, S.G. Veprev, V.K. Shumny, N.A. Kolchanov, S.E. Peltek, Vavilov Journal of Genetics and Breeding **17**, 765 (2013)
4. A.V. Obolenskaya, Z.P. Yelnitskaya, A.A. Leonovich, *Laboratory Works on Wood and Cellulose Chemistry* (EDP Ecology, Moscow, 1991)

Specialized pollinators and seed production of *Trollius* species introduced in Western Siberia

Anastasiya Gusar[1,*], and *Lubov* Buglova[2]

[1]FSEE HPE FSEI "Novosibirsk State Agrarian University", Russia
[2] FSIS "Central Siberian botanical garden SB RAS", Russia

Abstract. In five species of *Trollius* L., seed production of fruits was studied, plants being grown in culture in the forest steppe zone in Western Siberia. 2 species representing *Chiastocheta* genus were identified as specialized pollinators (*Diptera, Anthomyiidae*). Those insects pollinate the flowers and damage the seeds of three globeflower species: *T. apertus, T. europaeus T. asiaticus*. The flower season of other two species (*T. chinensis* and *T. ledebourii*) starts later and the plants can't be exposed to invasion, as *Chiastocheta* flies are at their larval stage yet. Pollination of these species is performed by non-specialized agents. The lowest rate of seed production realization of about 20% is registered in *T. europaeus*, while in *T. ledebourii* it is the highest comprising 71%. We consider the flies of *Chiastocheta* genus as an important part of the complex of factors lowering seed productivity in *Trollius* and their relationship as parasitism. Thus, in West Siberia for *Trollius* seed production increase, measures intended to reduce *Chiastocheta* abundance will be useful.

1 Introduction

Trollius L. representatives growing in temperate latitudes are frost-resistant and promising for introduction in Siberian region. They possess valuable decorative and medicinal properties [1, 2]. In *T. chinensis*, there have been extracted flavonoids exposing proved cancer-fighting activity [3].

Plants of *Trollius* genus being allogamous, pollination agents study turns to be a question of strategy for obtaining seeds. Flowering of introduced species shows its species-specific character in culture as well [4].

Specialized pollinators for *Trollius* genus the flies of *Chiastocheta* genus (*Diptera, Anthomyiidae*) are [5]. The influence of those agents is controversial. *Chiastocheta* imago individuals eat nectar of *Trollius* flowers and concurrently perform pollination gaining to production of seeds. On the other hand, larvae are parasites feeding on nothing but seeds and fruit tissues of *Trollius* genus plants. That decreases seed productivity [6, 7].

For *T. europaeus* L., *Chiastocheta* flies are described to be the only pollinator, as the closed globe-shaped flower makes getting in possible for small-sized insects exclusively. *T.*

[*] Corresponding author: gusara663@gmail.com

europaeus and *Chiastocheta* have mutual evolution, so their relationship can be identified as obligate mutualism. However, for other *Trollius* species *Trollius–Chiastocheta* interaction is parasitism, since the plants can also be pollinated by non-specialized agents as bees and bumblebees. Thus, the importance of specialized agents for seed production in species with open flower turns less sufficient [8].

The study targets to identify *Chiastocheta* species and to assess seed production in their host-plants of *Trollius* genus from the collection of Central Siberian botanical garden SB RAS.

2 Materials and Methods

The collection ground of Central Siberian botanical garden is located in the south-east of the West Siberian Plain in the southern outskirts of Novosibirsk. The coordinates are N 54.82; E 83.10 by google/maps; h: 130m.a.s.l. seed production in five species was assessed in 2017.

Chiastocheta species were identified in the generally recognized way according to the key reference-books [9, 10] by male genitals morphology (surstyli, cerci and V sternite projections) and by comparison with the identified species of St. Petersburg outskirts from the collection of Zoology Institute RAS.

Chiastocheta eggs and larvae along with *Trollius* seed production were calculated per fruit comprising a number of uncombed follicles (Table 1, 2). Seed productivity was studied according to the technique proposed by I.V.Vaynaghy [11] and being modified for aggregate follicles of *Trollius*. The percentage of seeds eaten by larvae was calculated from the total number of the developed ones. Potential seed productivity was assessed with a *SteREO Discovery V12* stereomicroscope photos taken by *AxioCam MRc-5* camera.

3 Results

For all the *Trollius* species, we have determined *Diptera* и *Hymenoptera* orders representatives as pollinators. In every *T. Europaeus* individual, sepals are not completely closed in the upper part, thus the flowers can be pollinated by non-specialized agents (Fig. 1).

Fig. 1. *T. europaeus* in USU № 440534 collection.

In *T. apertus* and *T. europaeus*, the terms of flowering are very similar, being mainly overlapped by flowering of the aboriginal *T. asiaticus* species. For them, two specialized pollinator species have been revealed: *Ch. trollii* and *Ch. sp.* (*rotundiventris*); both got acclimatized to the introduced successfully.

In *T. chinensis* and *T. ledebourii* the flowering starts about three weeks later than in *T. asiaticus*. During this period, all the specialized flies are passing the larvae stage; that is why summer-blossoming species can't de invaded. Larvae emergence from eggs occurs on the 2–4th day after the flowering period of the vulnerable species is over.

Table 1. Eggs and larvae number on *Trollius* fruits.

Number / Species	N		Eggs		Larvae	
	total	Without larvae	maximum	average	maximum	average
T. apertus	6	1	8	2.83	6	2
T. asiaticus	15	4	6	1.53	3	0,67
T. europaeus	14	3	5	1.93	3	1,8

Number of eggs blown by *Chiastocheta* flies onto the host plant follicules can vary a lot (Table 1). The most active blowing is observed on *T. apertus* plants. There may be up to 8 eggs per fruit. Correspondingly, up to 6 larvae may be feeding on one fruit. However, in all the species, some fruits have been detected without any eggs or larvae. The largest number of totally larvae-free fruits is registered in *T. asiaticus*, thus providing a lower average rate of invasion. The average invasion rate in fruits of the aboriginal species turns to be more than three times lower, in comparison to the one in the introduced *T. apertus* и *T. europaeus*.

Table 2. Seed production of the globeflower introduced species.

Data to compare / Species	Follicle, actual average number of seeds	Fruit (aggregate follicle)		Developed seeds, %	Eaten seeds, %
		Actual average number of seeds	Potential average number of seeds		
T. apertus	10.23±0.24	63.67±18.22	226.03±39.98	28.38	19.62
T. asiaticus	9.90±0.29	66.67±8.31	224.88±16.88	30.23	17.93
T. europaeus	7.83±0.60	47.70±8.94	231.97±21.62	20.48	17.56
T. chinensis	13.10±0.16	94.06±7.60	316.48±12.96	32.31	-
T.ledebourii	11.14±0.21	216.09±17.93	324.62±17.41	71.48	-

The potential seed production of fruits is higher in the species not exposed to *Chiastocheta* invasion than in the vulnerable ones (Table 2). The actual seed production appears to be the highest in *T. ledebourii*, while the lowest – in *T. europaeus* up to complete withering of fruits during their maturation period. However, no direct dependence of productivity on the invasion rate is revealed. In *T. ledebourii*, the actual seed production is more than twice as high as in *T. chinensis*. Similar tendency is observed in the vulnerable species. All the three species involved have nearly equal potential production, while the actual seed production is the lowest in *T. europaeus*, despite the flyblown of *T. europaeus* and *T. asiaticus* plants being close.

4 Discussion

In the region under the study in West Siberia, we have detected 2 species of specialized pollinators *Ch. trollii* and *Ch. sp.* (*rotundiventris*), which are found in Europe as well [8]. Their behavior corresponds to the data of literature [12].

The tendency of the introduced *T. apertus* and *T. europaeus* species to higher invasion rate can most probably be explained by the peculiarities of *Trollius–Chiastocheta* consortium mutual evolution in each region, thus the aboriginal species *T. asiaticus* turns to be a little more resistant to the invasion by flies. This confirms the reports on plants ability to increase the content of *Chiastocheta*-protective compounds [13, 14].

Aside from the pollination agent's blown, seed production decrease in *T. europaeus* can additionally be caused by the introduction of plants to more severe climatic conditions.

Introduction effectiveness is determined by the plant ability to reproduction [15], so in West Siberia, *T. ledebourii* proved to be the most successful and productive, as here this species is affected by two more beneficial factors (rhythm and climate) than within its natural habitat. Both factors provide them the invulnerability to *Chiastocheta* larvae invasion. Pollination by non-specialized agents turns to be enough for *Trollius* species mass seed production when grown in culture under conditions of West Siberia.

In this connection, *Trollius–Chiastocheta* relationship should be considered as the form of parasitism even for *T. europaeus*. For Trollius species cultivation, measures intended to reduce *Chiastocheta* abundance are desired to increase plant seed production.

Authors are grateful to E.P. Narchuk for identified insects being put at our disposal to confirm the identification of *Chiastocheta* species under the study.

References

1. N. V. Shipchinsky. *Flora of the USSR* **7** (Moscow, 1937) In Russ.
2. M. N. Povydysh, N. V. Petrovva, N. I. Medvedeva Wild flowering plants, their component composition and biological activity **1**, 24-66 (2008) In Russ.
3. S. Wang, Q. Tian., F. An, Oncology Letters **12**, 1705-1710 (2016)
4. L. V. Buglova, Vestnik IrGSCHA **44**, 35-41 (2011) In Russ.
5. V. Michelsen, *A revision of the Anthomyiidae (Diptera) described by J. W. Zetterstedt*, Steenstrupia.37-65 (1985)
6. N. Jaeger, L. Despres, C. R. Acad. Sci. Paris **321**, 789-796 (1998)
7. L. Despres, N. Jaeger. J .of Evol. Biol. **12**, 822-831 (1999)
8. O. Pellmyr, Biol. J. of the Linn. Soc. **47**, 337-365 (1992).
9. W. Henning, Beitr. Entomol. **3**, 655-668 (1953)
10. J. E. Collin, Proc. R. Entomol. Soc. **23**:95-102 (1954)
11. I. V. Vaijganii, Bot. J.**59**, 826–830 (1974)
12. F. Pompanon, E. Pettex, L. Després, Acta Oecologica **29**, 233-240 (2006)
13. C. Gallet, S. Ibanez, L. Zinger, J. Chem. Ecol. **33**, 2078-2089 (2007)
14. S. Ibanez, C. Gallet, F. Dommanget, BMC Evol. Biol. **9**, 261 (2009)
15. R. A. Kapisonova. *Herbaceous plants of the USSR broad-leaved forests: ecologo-floristic and introduction characteristics* (Nauka, Moscow, 1985) In Russ.

BIO Web of Conferences **11**, 00019 (2018) https://doi.org/10.1051/bioconf/20181100019
Prospects of Development and Challenges of Modern Botany

Features of development of the clonal plant *Scutellaria scordiifolia* Fischer ex Schrank (*Lamiaceae*) in Siberia

Alexandra Guseva [*]

Central Siberian Botanical Garden SB RAS, Novosibirsk, Russia

Abstract. Ontomorphogenesis of individuals of the clonal stoloniferous vegetative annual plant *Scutellaria scordiifolia* was studied in natural habitats. Ramets are represented by partial dicyclic shoots with an adventious root system. The ontogenesis of *S. scordiifolia* can be expressed by a sequence of phases of morphogenesis: [a partial shoot – a branched partial shoot] (v–g3) – a system of partial shoots (g2–g3) – a partial shoot (g2–ss). Individuals annually die off completely, regeneration takes place due to formation of tuberous stolons which are formed on the maternal stolon.

1 Introduction

Currently, vegetatively propagated plants called clonal [1,2]. Clonal stoloniferous herbaceous polycarpous plants represent an isolated group of plants integrated on the basis of their common ability of vegetative propagation with the help of specific shoots. Development of stolons provides an intensive vegetative propagation of plants and better assimilation of environmental resources [3]. Significant separation of daughter individuals from the maternal one helps to remove competitive relationships among them. At present of great interest are vegetative short-lived plants which G. N. Vysotskiy described [4] as plants whose maternal specimens that gave a new vegetatively–daughter generation die off before long. Their ultimate forms are vegetative annuals, vegetatively-daughter plants of which exist not more than one year [5]. Study of clonal plants allowed Yo. L. Lyubarskiy [6] to set aside an additional branch in the evolution of life forms of angiosperms, it leads from herbaceous perennial plants through vegetative short-lived plants to vegetative annuals. The species under study *Scutellaria scordiifolia,* a clonal stoloniferous herbaceous polycarpous plant is a vegetative annual. The scientists note that the group of vegetative annuals is the least studied from the biomorphological point of view [7,8,9]. In this regard the aim of the work was study of features of individual development of the clonal vegetative annual *S. scordiifolia* in natural habitats.

2 Materials and methods

[*] Corresponding author: guseva.sc@list.ru

Ontomorphogenesis was described on the basis of study of *S. scordiifolia* individuals in the Republics of Khakasiya, Buryatia, Altai and Zabaikal Krai. Study was conducted on individuals in a number of communities: steppe meadow – true steppe –meadow steppe.

When determining a life form, I. G. Serebryakov's [8] ecological-morphological classification of life forms and G N. Vysotskiy's [4] classification based on vegetative mobility pattern and ways of vegetative propagation of plants were used. The notions of morphogenesis phases were also used [10,11,12]. Ontogenesis was studied in accordance with the concept of discrete description proposed by T.A. Rabotnov [13] and A. A. Uranov [14].

3 Results

In all studied habitats only one life form – stoloniferous is formed in *S. scordiifolia* individual. Sterilization and loss of propagation by seeds take place in most of clonal stoloniferous plants [15, 16]. Predominance of vegetative propagation is characteristic of *S. scordiifolia*, seed individuals have not been found in nature. In this regard, morphogenesis of only vegetative individuals (ramets) was studied. In ontogenesis they pass the following phases: [a partial shoot – a branched partial shoot] (v–g3) – a system of partial shoots (g2–g3) – a partial shoot (g2–ss).

Development of individuals at different ontogenetic stages occurs according to one scheme. Elongated hypogeonous stolons of order n+1 develop on the stolon of the maternal ramet from the lateral buds. Due to accumulation of nutrient substances in the internodes, they thicken almost completely (only one-two metamers in the basal part remain not thickened) and resemble a tuber. In scaly-leaf axils lateral buds which as a terminal one are buds of regeneration are set on the stolon. Stolons may branch out due to bud opening in the basal part of a tuberous section, these stolons of order n+2 are elongated and completely thickened. Secondary roots develop in the nodes over the total length of stolons. At the end of the vegetative season not thickened metamers of the stolons of order n+1 and internode in the basal part of stolons of order n+2 decay, each stolon with the secondary root system keeps existing independently. In the next year due to pulling approximate metamers, in the apical part of the stolon is formed an apogeotropic section, and from the terminal bud develops an above-ground orthotropic elongated annual shoot with a shortened basal part consisting of 2–5 approximate metamers with scale shaped leaves. The shoot branches out: in the above-ground sphere from lateral buds develop vegetative enrichment shoots and paracladia, in the underground sphere on the stolon are formed daughter stolons giving rise to a new ramet generation. The maternal stolon gradually dies off from the basal end during a year. Thus, *S. scordiifolia* exists in the form of partial dicyclic shoots.

Ontogenesis of ramets begins with the virginal stage (Fig. 1). A partial shoot is vegetative, in late June it begins branching out: in the above-ground part develop elongated orthotropic vegetative shoots, and on the stolon start in growth stolons of order n+1. In the next year new ramets would be at the virginal and young generative stages.

Virginal individuals with reduced vitality may be also formed in the populations. Such individuals are represented by monocyclic anisotropic shoots with adventious root system, they are formed from the lateral buds on the plagiotropic section of the stolon of the individuals at the middle-aged generative stage, completely die off at the end of the vegetative season.

At the young generative stage a partial shoot is generative, an inflorescence - simple frondose raceme. As a result of branching in the above-ground part of a shoot develop elongated vegetative enrichment shoots, and on the stolon emerge new thickened stolons coloured violet from the distal end. In late vegetative season the maternal individual dies off, and stolons overwinter. Ramets pass to the middle-aged generative stage.

At the middle-aged generative stage on the tuberous stolon on the apogeotropic part in the axils of 1–2 metamers serial buds are set, there are only two of them: lower and upper. The lower bud is set later, but in the process of development it overgrows the upper one. Stolon of order n+1 developing from the lower bud is more vigorous, may reach 30 cm in length, 0,3–0,5 cm in diameter, at most branches out, stolons of order n+2 appear. On the base of stolons of order n+1 develop middle-aged and old generative individuals, and on the base of stolons of order n+2 – virginal and young generative ones. Stolons of order n+1 developing from upper serial buds and usual lateral buds do not branch out, on the base of them develop middle-aged generative individuals. In total from 5 to 10 (20) new stolons develop on the maternal stolon.

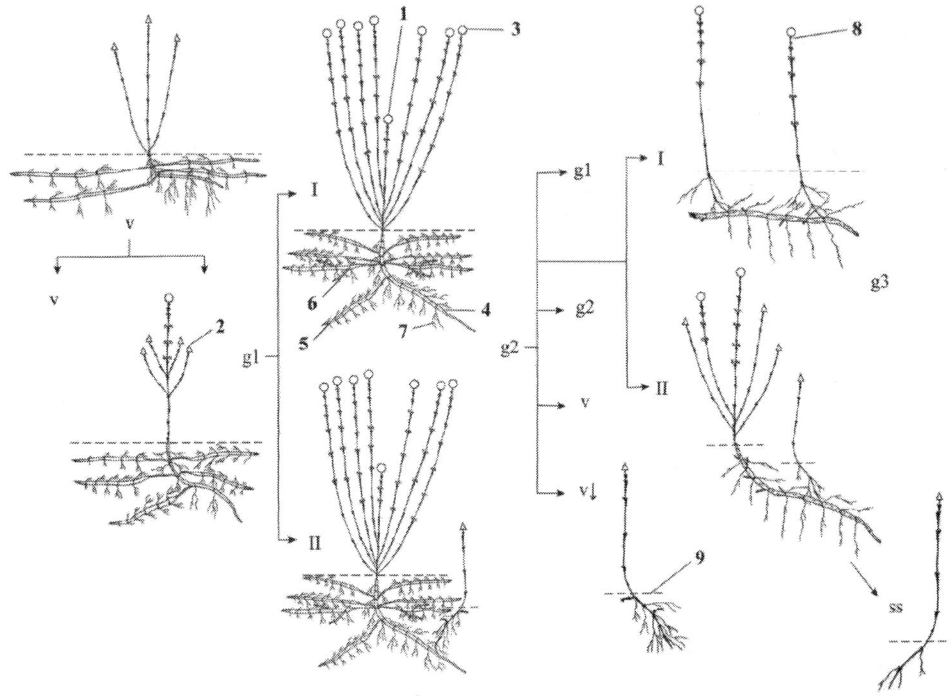

Figure 1. Ontomorphogenesis of *Scutellaria scordiifolia* stoloniferous life form.
Legend – I, II – variants of individual structures at the ontogenetic stage; v↓ – an individual at the virginal stage with reduced vitality. 1 – a dicyclic anisotropic generative shoot, 2 – a monocyclic vegetative enrichment shoot, 3 – paracladium, 4 – a tuberous stolon of order n, 5 –a tuberous stolon of order n+1, 6 – a stolon of order n+2, 7 – adventitious roots, 8 – a monocyclic orthotropic generative shoot, 9 –soil level.

The tuberous stolon of the individuals at the old generative stage is covered by transversal and longitudinal constrictions and has a yellow tint. Most often at the old generative stage new stolons do not develop or die off at the early stages of development and individual ontogenesis is over. In some cases lateral buds on the plagiotropic part of the stolon start in growth, a system of partial shoots is formed. In the same year the system disintegrates, a shoot decays at the base and separates from the maternal stolon, emerges a ramet at the subsenile stage. Such an individual dies off completely in late vegetative season.

Sometimes a terminal stolon bud at the old generative stage dies off after overwintering, then lateral buds on the plagiotropic stolon part start in growth, a system of partial shoots, which does not disintegrate as a rule, is formed. Such structures may also form at middle-aged generative stage, monocyclic anisotropic vegetative shoots develop from the lateral buds. In these cases the disintegration of the system occurs due to decay of the basal part of a shoot, new ramets at the virginal stage are formed . We consider such ramets to belong to the virginal individuals with reduced vitability.

Thus, *S. scordiifolia* throughout the range in different conditions of habitation develops as a clonal stoloniferous herbaceous polycarpous plant. It is a vegetative annual according to the classification of G. N. Vysotskiy. As in most of stoloniferous species prevails vegetative propagation which takes place due to development of tuberous branched stolons, at the middle-aged generative stage the intensity of vegetative propagation increases due to opening of serial buds. Ramets are represented by dicyclic partial anisotropic shoots with an adventious root system. Each vegetatively-daughter generation dies off completely at the end of the vegetative year. The course of ontogenesis of *S. scordiifolia* is similar to the ontogenesis of species with a similar life form. Ontogenesis is abbreviated and represented by a cycle of development of specialized dicyclic anisotropic shoots. Particularization of individuals can lead to a rejuvenation of the daughter particles to the virgin stage. Depending on the individual ontogenetic state through the following phases of morphogenesis: [a partial shoot – a branched partial shoot] (v–g3) – a system of partial shoots (g2–g3) – a partial shoot (g2–ss).

The work was carried out with the financial support of the grant of RFBR within the framework of scientific project № 18-04-00621-a and project of the State Assignment of Central Siberian Botanical Garden of the Siberian Branch of the Russian Academy of Sciences № AAAA-A17-117012610053-9.

References

1. J.C. Noble, A.D. Bell, J.L. Harper, J. of Ecology **67**, 983-1008 (1979)
2. P.K. Diggle, S. Lower, T.A. Ranker, Int. J. of Plant Sciences. **159**, 606-615 (1998)
3. V.C. Terry, A.C. Bengt, S.J. Ingibjörg, M.S. Brita and S. Jonasson, Oikos. **63**, 341-347 (1992)
4. G.N. Vysotsky, Bul. of applied Bot. **8**, 1-330 (1915)
5. E.L. Lyubarsky, Bot. J. **46**, 959-968 (1961)
6. E.L. Lyubarsky, Bul. Bot. Gard. SSU. **5**, 362-365 (2006)
7. Yo. N. Berko, *Abstracts of VII UBS Congress* (Kiev, 1982)
8. I.G. Serebryakov, *Ecological morphology of plants* (Sov. Nauka, Moscow, 1962)
9. O.A. Korovkin, *On the regularities of the ontogeny of a clone using the example of stoloniferous herbaceous polycarpous* (Moscow, 1999)
10. O.V. Smirnova, L.B. Zaugol'nova, *Cenopopulation of plants* (Nauka, Moscow, 1976)
11. V.A. Cheryomushkina, *Biology of Allium species in Evrasia* (Nauka, Novosibirsk 2004)
12. N.P. Savinykh, V.A. Cheryomushkina, Cont. Prob. of Ecology. **8**, 541-549 (2015)
13. T.A. Rabotnov, Tr. BIN AN SSSR, Ser. 3. Geobotanika. **6**, 179-196 (1950)
14. A.A. Uranov, Naych. Dok. Vysshaya Shkola. **2**, 7-34 (1975)
15. C.G. Eckert, Evolutionary Ecology. **15**, 501-520 (2002)
16. T. Herben, B. Šerá, J. Klimešová, Oikos. **124**, 469-476 (2015)
17. O.A. Korovkin, Izv. of TAA. **6**, 57-67 (2013)
18. E.A. Alyabysheva, *Ontogeny and features of the organization of cenopopulations of some hygrophytes of the Republic of Mari El* (Yoshkar-Ola, 2001)

Fir-spruce and lime-fir-spruce forests of Vyatka-Kama biome

Nikita G. Kadetov[1,*]

[1]Lomonosov Moscow State Universty, Faculty of Geography, GSP-1, 1 Leninskiye Gory, 119991, Moscow, Russia

Abstract. During the preparation of map "Biomes of Russia" in broad-leaved-coniferous forests belt of the Eastern Europ Vyatka-Kama biome was substantiated. This territory is allocated for the first time at such a high level. Formations of fir-spruce and lime-fir-spruce forests occupy a central position in the vegetation cover of the biome and largely determine its specificity. As the result of ecological-morphological classification 7 groups of associations and 17 associations of vegetation have been determined.

1 Introduction

Broad-leaved-coniferous forests of Eastern Europe belong to the most controversial in their rank and attribution to a particular chorologic units. Using the biome concept of biosphere organization the map "Biomes of Russia" was developed in the department of biogeography of Lomonosov Moscow State University [1]. During map preparation and justification of biogeographic boundaries in it, special attention was paid to the hemiboreal forests of European Russia. Based on research [2] two biomes – Smolensk-Privolzhsky and Vyatka-Kama – were substantiated in broad-leaved-coniferous forests belt of the European part of Russia. Key indicators for establishing their boundaries were: biotic diversity - floristic (faunistic) and cenotic - and bioclimatic aspects (mean annual temperatures, sum of biologically active temperatures > 10 °C and average annual precipitation).

The Vyatka-Kama biome is situated in the middle Volga basin and its largest left tributary, the Kama. Distinguished in it are the Zavolzhie rudimentary elevation, the Vyatka Uval and the Vyatka-Kama interfluve in the east (parts of the Nizhny Novgorod and Kirov regions and the Republics of Mari El, Tatarstan and Udmurtia) and the left bank of the Kama with the Tulva Upland and the spurs of the Ufa Plateau (parts of Perm region, Udmurtia, Bashkortostan, Sverdlovskaya region). This territory is allocated for the first time at such a high level.

Woodland coverage of Vyatka-Kama biome reaches 51%, but differs significantly in the western (40-45%) and eastern parts (60-65%). Among forests, the fir-spruce (*Abies sibirica* Ledeb., *Picea* x *fennica* (Regel) Kom., *P. obovata* Ledeb.) and lime-fir-spruce (*Tilia cordata* Mill.) forests and their derivatives, which account for more than 83% of the wooded area, are absolutely dominant in the area occupied. Lime-fir-spruce forests are to a

[*] Corresponding author: biogeonk@mail.ru

certain extent unique for this biome and in many respects determine its specificity. At the same time, almost 60% of the territory falls on derivative forests with a predominance of birch (*Betula pendula* Roth., *B. alba* L.) and aspen (*Populus tremula* L.). Spruce (1.2%) and broadleaf-spruce (1.8%), pine (*Pinus sylvestris* L.) (5.4%) and broadleaf-pine (3.9%), oak (*Quercus robur* L.) (0,7%) and lime (2.5%) forests are also presented. Not more than 2% of the territory is occupied by bogs.

2 Materials and methods

Formations of fir-spruce and lime-fir-spruce forests occupy a central position in the vegetation cover of the biome and largely determine its specificity. As a part of the work to justify the boundaries of biomes using the ecological-morphological classification approach [3, 4], an assessment of the diversity and distribution of fir-spruce and lime-fir-spruce forests was made. It was based on the original field research materials (more than 100 plots), vegetation maps and published data. As the result 7 groups of associations and 17 associations of vegetation have been determined.

3 Results and discussions

3.1 Fir-spruce forests

Fir-spruce forests are widespread in the biome. They occur from the northern part of the Vetluga basin to the eastern boundary of the study area, along the watersheds and on the slopes of river valleys on sandy loamy and loamy podzolic and sod-podzolic soils. Fir is located here near the western limit of its distribution. In the west, it does not form pure forests – the predominant breed is spruce, and the participation of fir does not exceed 0.1-0.4 of stand composition. To the east its share increases up to 0.3-0.6 (0.7), while the fir-spruce forests themselves begin to predominate over the spruce forests. At the same time, the consideration of fir-spruce forests apart from spruce forests is due, first of all, to the fact that together with fir in the addition of these forests, a number of siberian species are essential. In the area of research fir-spruce forests are one of the leaders in the formation of forest cover, and the territory of the Priuralie, along with the Pechora basin, is the main arena of their spread. Also, similar types are known from the Urals and the extreme west of Siberia [5, 6].

Fir-spruce greenmoss forests - the most widespread group of fir-spruce forests of the Priuralie, the most represented in its northern half. They are confined to territories with a predominance of moraine and eluvial-deluvial loams and clays with sod-strongly podzolic and podzolic soils.

As part of the stand, in addition to spruce and fir, there is an admixture of birch, aspen, lime, less often – pine. Oak and lime are noted in the regrowth. In a rare undergrowth are represented: *Frangula alnus* Mill., *Sorbus aucuparia* L., *Juniperus communis* L., and also *Lonicera caerulea* L. In the grass and shrub layer, the followings are common: *Maianthemum bifolium* (L.) F. W. Schmidt, *Trientalis europaea* L., *Vaccinium myrtillus* L, *V. vitis-idaeae* L., *Oxalis acetosella* L., *Pyrola rotundifolia* L., *P. minor* L., *Gymnocarpium dryopteris* (L.) Newm., *Carex rhizina* Blytt ex Lindbl.. In a sparse moss cover, *Pleurozium schreberi* (Brid.) Mitt., *Hylocomium splendens* (Hedw.) Bruch et al., *Dicranum* sp., etc. predominate.

Fir-spruce haircup-moss (*Polytrichum commune*) forests are spread slightly, mostly to the west of Vyatka, and occupy a variety of depressions with poor drainage and sod-podzolic gley soils. In their stand, the role of birch and pine is quite noticeable. The herbaceous and shrub layer is usually very sparse and is represented by: *Vaccinium myrtillus*, *Equisetum sylvaticum* L., *Molinia caerulea* (L.) Moench, *Carex nigra* (L.) Reichard, etc. The moss cover is dominated by *Polytrichum commune* Hedw.

Fir-spruce grass forests are occupied by a few swamped old valleys in the floodplains of small rivers and the bottoms of ravines. The stand of the alder (*Alnus glutinosa* (L.) Gaertn) and birch is inherently significant. Characteristic is the participation in the undergrowth of species of Euro-Siberian distribution – *Lonicera caerulea* and *Swida alba* (L.) Opiz, as well as *Atragene sibirica* L.. In the high grass stands there are: *Filipendula ulmaria* (L.) Maxim., *Valeriana officinalis* L., *Cirsium palustre* (L.) Scop., *Matteuccia struthiopteris* (L.) Todaro, *Glyceria maxima* (C. Hartm.) Holmb., *Cacalia hastata* L., etc.

Fir-spruce nemoral forests are confined to watershed spaces with the richest (in particular, sod-carbonate) soils (in the east of the biome), as well as parts of river valleys (mainly in the west of the biome). This is one of the most important groups that characterize the vegetation cover of the whole region. The stand is characterized by a noticeable admixture of birch and aspen and a pronounced second tier of lime with the participation of maple (*Acer platanoides* L.) and elm (*Ulmus glabra* Huds.). In the undergrowth sometimes there is abundant *Sorbus aucuparia*, but mostly it comprised of nemoral species – *Lonicera xylosteum* L, *Corylus avellana* L., *Euonymus verrucosa* Scop.. For the grass and shrub layer is characterized by the participation of boreal, including Siberian, species (*Oxalis acetosella*, *Linnaea borealis* L., *Carex digitata* L., *Trientalis europaea*, *Aconitum septentrionale* Koelle, etc.) and nemoral Euro-Siberian species (*Aegopodium podagraria* L., *Pulmonaria obscura* Dumort., *Stellaria holostea* L., *Galium odoratum* (L.) Scop., *Paris quadrifolia* L., etc.).

Various authors [7, 8] repeatedly stressed complex and multifaceted nature of the vertical structure of this forests in which an alternation in tiers formed species having different origin. Thus, the tier of European nemoral species alternates with the tier of Euro-Siberian taiga species. All this indicates the "borderline" nature of these forests, which in their way are "transitional" from the East European to the Priuralie and the Urals.

3.2 Lime-fir-spruce forests

Lime-fir-spruce forests are one of the outstanding features of this biome. They are predominant over the Kama - in Priuralie. Significant massifs are presented in the Vyatka-Kama interfluve. Most of these forests are associated with mesopodzolic, rarely - gray forest soils formed on eluvial-diluvial loams and clays indigenous Permian rocks. Brought forth near the extreme north-eastern distribution limits of broad-leaved trees and their companions, these forests are characterized by an increase in the composition of all the layers of the role of Siberian species and, in part, lime, reduced the role of a maple, an increase of cover herbs (especially - tall and ferns) and a decrease in participation shrubs.

The composition and structure of these forests are often significantly transformed, mainly by logging. In large areas of the forest reduced at all or replaced by birch, aspen and (significantly less) lime forests.

Lime-fir-spruce high-grass forests are prevalent mainly in the northern part of the Priuralie. They are mainly confined to deep podzolic heavy loamy and brown clay soils of the upper parts of the slopes of hills and watersheds. The stand is composed of spruce, fir, and inferior in height lime, sometimes with an admixture of elm. The undergrowth is rare from *Lonicera xylosteum*, *Frangula alnus*, *Sorbus aucuparia*. Grass cover consists of *Aconitum septentrionale* (almost always - a clear dominant), *Dryopteris expansa* (C. Presl.) Fraser-Jenkins et Jermy, *Oxalis acetosella*, *Dryopteris filix-mas* (L.) Schott, *Stellaria holostea*, *S. bungeana* Fenzl., *Maianthemum bifolium*, etc. In a few sparse moss cover is dominated *Pleurozium schreberi*, *Brachythecium reflexum* (Starke) Bruch et al., etc.

Lime-fir-spruce large-fern forests are confined to the slopes of watersheds and the bottoms of ravines with deeply podzolic heavy loamy soils. The stand is characterized by smaller participation of fir in comparison with the previous group, an admixture of birch and aspen, rarely – pine. *Sorbus aucuparia, Lonicera xylosteum* and *L. caerulea* are in the rare undergrowth. Quite a dense grass layer is formed by *Dryopteris expansa, Diplazium sibiricum* (Turcz. ex G. Kunze) Kurata, *Athyrium filix-femina* (L.) Roth, *Calamagrostis obtusata* Trin., *Oxalis acetosella, Stellaria bungeana*, etc. The moss cover is dominated by *Hylocomium splendens, Brachythecium reflexum.*

Lime-fir-spruce nemoral forests are the most common in the biome. They predominate in the area in the Priuralie and occupy significant positions in the Vyatka-Kama interfluve. In general, they are associated with watershed spaces and floodplain slopes with sod-podzolic or brown clayey and loamy soils. The stand is characterized by an increase in the role of fir and lime, often noticeable admixture of elm of rough and aspen, rarely – maple. In the undergrowth are represented *Sorbus aucuparia, Lonicera xylosteum, Viburnum opulus* L. For the grass and shrub layer the common species are *Aegopodium podagraria, Asarum europaeum* L., *Stellaria holostea, S. bungeana, Brachypodium sylvaticum* (Huds.) Beauv., *Galium odoratum, Carex rhizina, Lathyrus vernus* (L.) Bernh., *Aconitum septentrionale, Cicerbita uralensis* (Rouy) Beauv., etc. The moss cover is usually poorly expressed.

References

1. G.N. Ogureeva (ed.), *Biomes of Russia. Map M 1: 7 500 000* ("Financial and Organizational Consulting", Moscow, 2016)
2. N.G. Kadetov, Izv. Samarsk. Nauchn. Centra RAN, **14**, 1603 (2012)
3. E.M. Lavrenko, Botan. Zhurn., **67**, 572 (1982)
4. A.P. Shennikov, *Introduction to geobotany* (Izd-vo Leningr. Un-ta, Leningrad, 1964)
5. G.A. Voronov, L.M. Trofimova, S.V. Balandin, *Complex fir-spruce forests of the Ural Prikamye* (Izd. Perm. Un-ta, Perm, 2005)
6. L.P. Rysin, Yu.I. Man'ko, S.M. Bebiya, Fir forests of Russia (Tov-vo nauchn. izd. KMK, Moscow, 2012)
7. V.V. Alyokhin, *Explanatory note to the geobotanical maps (modern and restored) of the former Nizhny Novgorod province (scale 1: 500 000)* (Perv. Kartogr. f-ka VKT, Leningrad, 1935)
8. S.S. Stankov, *Essays on the physical geography of the Gorky region. 3rd ed., corr.* (Gork. Obl. Izd-vo, Gorky, 1951)

BIO Web of Conferences **11**, 00021 (2018) https://doi.org/10.1051/bioconf/20181100021
Prospects of Development and Challenges of Modern Botany

Phylogenetic relationships between *Chondrilla* L. species in European Russia based on plastid DNA sequencing

Alexandr Kashin[1,*], *Ivan* Schanzer[2], *Tatyana* Kritskaya[1], and *Alena* Parkhomenko[1]

[1]Saratov State University, Botanic Garden, Saratov, 410012, Russia
[2]Main Botanical Garden, Herbarium (MHA), Moscow, 127276, Russia

Abstract. The article examines the genetic diversity in 54 populations of 9 *Chondrilla* species (*C. acantholepis, C. ambigua, C. brevirostris, C. canescens, C. graminea, C. juncea, C. laticoronata, C. latifolia,* and *C. pauciflora*) in European Russia. Plastid DNA segments trnT–trnF are selected as markers. Reconstruction of evolutionary networks based on the principle of maximum parsimony reveals that the sample is divided into four groups, where group 1 is *C. ambigua*, 2 – *C. brevirostris*, 3 – *C. laticoronata*, 4 – *C. acantholepis, C. canescens, C. graminea, C. juncea,* and *C. latifolia*. The findings show that *C. acantholepis, C. canescens, C. graminea, C. juncea* and *C. latifolia* are to be treated as synonyms under the name of *C. juncea*.

1 Introduction

Chondrilla genus is comprised by approximately 30 species that form two subgeni and four sections [1]. The genus is wide spread in the steppes and deserts of Eurasia and North Africa. There are no less than 7 in vivo *Chondrilla* species in European Russia. Six of them – *C. acantholepis* Boiss., *C. brevirostris* Fisch. et Mey, *C. canescens* Kat. et Kir., *C. graminea* Bieb., *C. juncea* L., and *C. latifolia* Bieb.- belong to the section *Chondrilla* of the subgenus *Chondrilla*. The seventh species – *C. ambigua* Fisch. – belongs to the subgenus *Brachyrynchus* [1]. Lately, more *C. laticoronata* Leonova populations are being spotted [2, 3].

Currently, there is no consistency in representing the taxonomic structure of the *Chondrilla* genus. Apomictic reproduction that is highly characteristic of the *Chondrilla* species [4, 5] makes it problematic to single out the species and distinguish between them as well as causes contradiction in evaluating their rank and taxa's relationships. The embryological analysis and the analysis of seed productivity show that, out of all the studied species in European Russia, *C. ambigua* is the only obligate amphimictic species while all the remaining taxa (*C. acantholepis, C. canescens, C. graminea, C. juncea, C. latifolia, C. brevirostris,* and *C. laticoronata*) are facultative apomicts [3, 5].

* Corresponding author: kashinas2@yandex.ru

The present paper attempts to identify the relationships between the *Chondrilla* species in Southern Russia, based on the plastid DNA sequencing.

2 Materials and Methods

2.1 Sampling

The specimens were collected from the natural populations of 9 *Chondrilla* species (*C. acantholepis*, *C. ambigua*, *C. brevirostris*, *C. canescens*, *C. graminea*, *C. juncea*, *C. Laticoronata*, *C. latifolia*, and *C. pauciflora*) in the Astrakhan, Volgograd, Penza, Samara, Saratov, and Ulyunovsk oblasts as well as the Rebublic of Kalmykia and the Republic of Crimea. Three plants were selected from every population.

2.2 DNA Extraction and Sequencing

DNA was extracted from the ligules (petal-like corollas) after they had been dried in silica gel, using NucleoSpin® Plant II (MACHEREY-NAGEL, Germany) according to the manufacturer's protocol.

Primers *a* and *b*, *c* and *d*, *e* and *f* were used to perform the amplification of the plastid DNA segments *trnT-trnL* and *trnL-trnF*, including the intron *trnL* [6].

Sequencing was carried out in the ABI PRISM 3130 XL sequenator using the reagent kit BIG DYE TERMINATOR kit ver. 3.1 produced by the Sintol company (Moscow, Russia).

2.3 Statistical Analysis

The forward and reverse sequences were manually edited and aligned in BioEdit 7.0.5.3. program [7]. The haplotype extraction and building of rootless network employed the method of statistical parsimony in TCS v. 1.21 program [8]. Only the nucleotide recombinations were taken into account; the indels were considered missing data. In order to root the obtained network, the data were further subject to the Neighbor Joining analysis in the program SplitsTree4 ver. 4.11.3[9]. An outgroup was comprised of the DNA sequences of *Helenium arizonicum* S.F.Blake (DQ395174.1) and *Giallardia comosa* A.Gray (DQ395152.1) obtained from GenBank.

3 Results and Discussion

The associated alignment of the spacers *trnT-trnL* and *trnL-trnF* and the intron *trnL* included 1292 positions, 13 of which were variable (contained recombinations). Furthermore, the research identified the duplication of the TATTT fragment in all specimens excluding *C. ambigua*, *C. pauciflora,* and *C. brevirostris*; the deletion of GAAA fragment in all specimens except for *C. ambigua*, *C. pauciflora*, *C. brevirostris*, and *C. laticoronata*; and the duplication of ATTACAAAT fragment in all *C ambigua*, *C. brevirostris*, *C. pauciflora*, and *C. laticoronata* specimens. The analysis in the TCS program allowed for uniting *Chondrilla* sequences in 15 haplotypes. To reconstruct haplotype genealogy, all haplotypes were mapped in one network (probability of parsimonious decision >95%) that included 26 haplotypes, 11 of which were absent in the sample and were hypothetical transitional haplotypes (Fig. 1 on the *left*). Topologically, 4 clades, or groups were singled out within the network; each group was constituted of the internal haplotype and the closest terminal haplotypes designated with letters and numbers.

In order to root the network, the haplotypes' matrix was subject to the Neighbor Joining analysis; the sequences of *Helenium arizonicum* and *Giallardia comosa* performed as an outgroup. The topology of the philogentic tree and the topology of the network were completely corresponding to each other. However, the place of the hypothetical root remained undefined which, most likely, is explained by the absence of the haplotypes situated between the clade A and its sister clade constituted of all the remaining haplotypes.

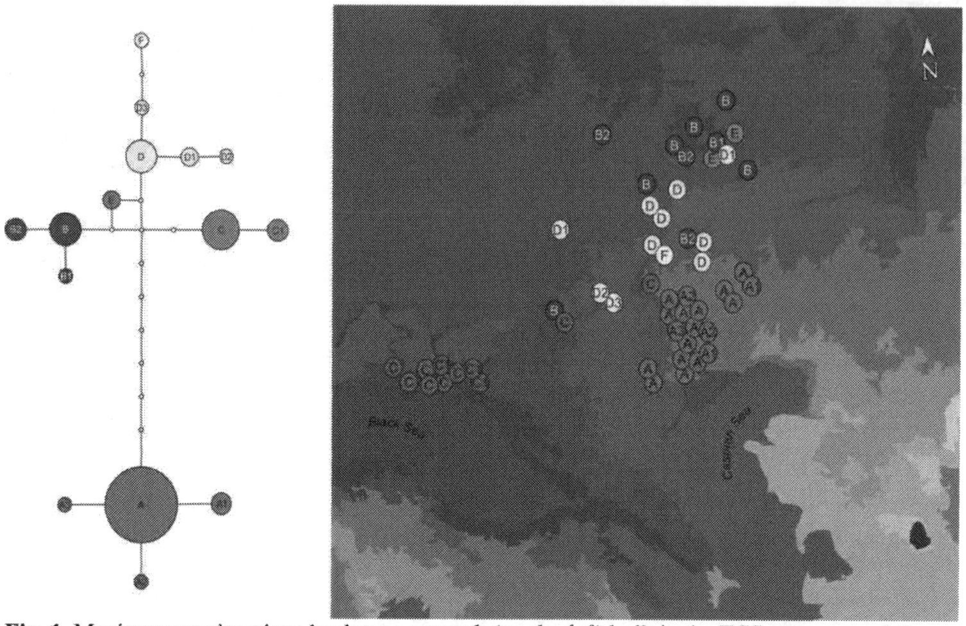

Fig. 1. Maximum parsimonious haplotype network (on the *left*) built in the TCS program and the locations of the studied *Chondrilla* populations (on the *right*). The circles' sizes are proportional to the numbers of specimens with a given haplotype. The small empty circles represent missing haplotypes.

Geographical mapping of haplotypes revealed that the haplotypes are distributed in the clear-cut patterns (Fig. 1 on the *right*). Group A is constituted of all specimens of *C. ambigua*, *C. brevirostris*, *C. laticoronata* and the probable hybrid species *C. brevirostris* × *ambigua,* located in the Astrakhan oblast and Kalmykia. Group A also contains specimens of *C. brevirostris* and *C. pauciflora* from Western Kazakhstan. Also, 4 out of 5 *C. ambigua* populations included specimens that possessed only haplotype A sequences. Haplotype A3 was found in *C. ambigua* population located near the Volnoye village, Kharabalinsky rayon, Astrakhan oblast. Apart from specimens of *C. ambigua*, haplotype A was found in 4 out of 5 *C. brevirostris* populations and in 1 out of 3 *C. laticoronata* populations. Haplotype A1 was found only in *C. brevirostis* population in Astrakhan oblast and in *C. laticoronata* population in Western Kazakhstan, haplotype A2 – only in *C. laticoronata* populations in Astrakhan oblast. No single instance of polymorphism within local populations was revealed.

Haplogroups B, C, D and E are widespread in *C. acantholepis*, *C. canescens*, *C. graminea*, *C. juncea*, and *C. latifolia* populations and are not confined to any particular taxon.

Both the analysis of chloroplast ITS sequences and the earlier conducted analysis of nuclear ribosomal ITS sequences [10] show that there are two evolutionary routes within the studied sample. The first route is observed in *C. ambigua* and *C. pauciflora*, as well as in *C. brevirostris* and *C. laticoronata* that have the plastid haplotypes of the same group

(A). The latter two species are characterized by high heterozygosity of the ITS sequences [10]. The second evolutionary route is represented by the species with the haplotype groups B-F. Based on ITS ribotypes, *C. laticoronata* and *C. brevirostris*, though possessing the same chloroplast haplotypes as *C. ambigua*, are drastically different from *C. ambigua* and completely heterozygous. Both facts provide strong evidence that *C. laticoronata* and *C. brevirostris* derived from the hybridization of *C. ambigua* with some other species that resulted in the formation of new stable apomictic clones.

In conclusion, there is no doubt that *C. ambigua* is an independent species. *C. brevirostris* and *C. laticoronata* are hybrid groups, the parental forms of which are missing in the sample. The obtained data prove that *C. acantholepis*, *C. canescens*, *C. graminea*, *C. juncea*, and *C. latifolia* are the specimens of one and the same species under the priority name of *C. juncea*.

This study was supported by the Russian Foundation for Basic Research (project #15-04-04087)

References

1. T. G. Leonova, *Chondrilla* L., In : *Flora of European Part of the USSR, Volume 8* (Nauka Publ., Leningrad, 1989)
2. P. F. Maevsky, *Flora of the Middle Belt of the European Part of Russia* (KMK Scientific Press Ltd., Moscow, 2014)
3. E. V. Ugolnikova, A. S. Kashin, A. O. Kondrateva, Izv. Sarat. Univ. Ser. Chem. Biol. Ecol. **17,** 53-61 (2017)
4. P. J. van Dijk, Phil. R. Soc. Lond. B. **358**, 1113-1121 (2003)
5. A. S. Kashin, A. O. Popova, I. S. Kochanova, E. V. Ugolnikova, Yu. A. Polyakova, Bot. Zhurnal **100,** 828-840 (2015)
6. P. Taberlet, L. Gielly, G. Pautou, J. Bouvet, Plant Mol. Biol. **17**, 1105-1109 (1991)
7. T. A. Hall, Nucl. Acids. Symp. Ser. **41**, 95-98 (1999)
8. M. Clement, D. Posada, K. A. Crandall, Mol. Ecol. **9,** 1657-1659 (2000)
9. D. H. Huson, D. Bryant, Mol. Biol. Evol. **23**, 254-267 (2006)
10. A. S. Kashin, T. A. Kritskaya, A. S. Parkhomenko, A. O. Kondratyeva, N. A. Petrova, *Proceedings of the Conference Dedicated to the 85th Anniversary of the Birth of V.N. Tikhomirov* (MAKS Press, Moscow, 2017)

Anatomical structure of leaves, stems and roots of hemiparasitic plants Thesium ebracteatum Hayne (Santalacea R. Br.) from the Urals

Olga Kiseleva[1,*], and *Ludmila* Trofimova[2]

[1]Russian Academy of Sciences, Ural Branch: Institute Botanic Garden, 620144, Ekaterinburg, Russia
[2] Ural Federal University, Ekaterinburg, 620000, Russia

Abstract. The paper presents the results of structure researches of *Thesium ebracteatum* Hayne. plants from the Ural populations. The anatomic description includes the characteristic of leaves and roots, the stems structure in its top, middle and lower part. Some stomatografic parameters of leaves are calculated. This is the fist anatomical report on the important diagnostics signs of *Thesium ebracteatum* Hayne plants.

1 Introduction

Detailed anatomical studies of *Thesium ebracteatum* Hayne plants from the Urals were not carried out. This species is known from sites in central Europe, European part of Russia, Western Siberia and in the Urals. It is a native herb of forest lawns and thickets of bushes. It is perennial hemiparasitic herb with a subterranean rhizome and numerous storage lateral root with haustorial connections. The aerial shoots annual, erect or ascending at the top, 10-30 cm tall, usually unbranched, glabrous, rarely leafy. At the top of the shoots presents bracts devoid of flowers; leaves linear, glabrous. Inflorescence is a simple, unbranched brush; flower on short stalk with one bract, bracts like the leaves but decreasing in size, perianth tubular, upward strongly wrapped. Capsule on the stalk is elliptical, with longitudinal branching veins, three times larger than the remaining part of the perianth. The paper presents the studies of anatomical structure of the leaf, stem and roots of *Thesium ebracteatum* plants from the Urals populations.

2 Materials and Methods

The plants were collected from the mountainous areas of Middle Ural (Sverdlovsk region) in two populations nearby Nizhnyaya Irga, Ust'-Bayak. Specimens for morphological study were dried according to standard herbarium techniques and stored in the Herbarium of Institute of Plant and Animal Ecology, Ural Branch of the Russian Academy of Sciences (SVER). The materials for anatomical study were preserved in 70% alcohol in the field. Anatomical observations were performed on transverse sections of root and stem (in its top,

* Corresponding author: kiselevaolga@inbox.ru

average and lower part) cut by rotary microtome Microm HM (Carl Zeiss, Germany), and surface sections of leaves cut by hand. All slides were observed by light microscopes Micros (Austria) and Zeiss Axioscopus (Carl Zeiss, Germany). Well-staining sections were photographed. Stomatal index (SI) has been calculated according to Salisbury (1927) [1]. All measurements and observations were made ten times on different slides. All microscopic measurements were made with the aid of an occular micrometer.

3 Results and Discussion

Leaf features of the species are shown in Figure 1. In surface preparations, the epidermal cells are covered by a thick cuticular layer. The leaf is amphystomatic with oval paracytic stomata. Epidermal cell walls are not undulate. The cells of the epidermis are isodiametric and polygonal with straight walls except in the coastal regions where they are narrow, elongated parallel to the veins. Upper epidermal cells have the same size as the lower ones; trichomes are absent, leaf margin with sharp papillae. Stomata orientation was random. Guard cells are elliptical, pores are elongated. Average size of guard cells in microns is 31 in lengths (varies from 16 to 38), 23 in widths (varies from 14, to 35). SI adaxial epidermis 12,8 and abaxial epidermis 11,3. The presens of stomata in adaxial and abaxial epidermis in equal parts may be the adaptive advantages to hemiparasitic nutrition. Lamina is dorsiventral. Mesophyll consists of 1 layer of elongated palisade cells and 2 or 3 layers of isodiametric, spongy parenchymatic cells have small intercellular cavities.

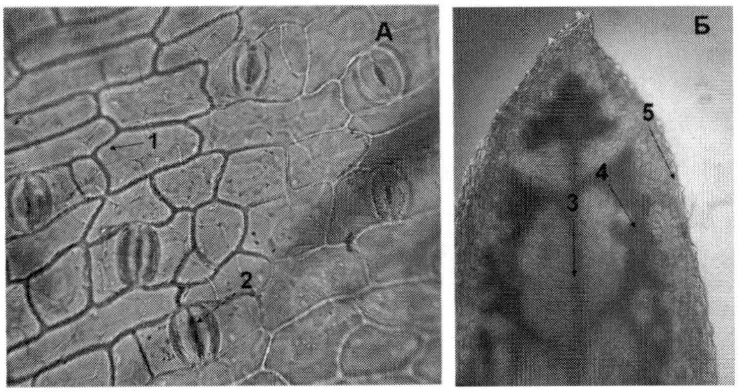

Fig. 1. Anatomical features of *Thesium ebracteatum* leaves. A Lower epidermis X400, Б Surface section of leaf X80. 1 – epidermis cell; 2 – stoma; 3 – Midrib; 4 – vascular bundle; 5 – leaf margin with sharp papillae.

A transverse section taken from the top, middle and lower part of the stem was observed (Fig. 2, 3A). The epidermal cells are covered by cuticular layer, trichomes are absent. Stem cortex consists of 2-5 layers of usually oval cells. Some cortex cells contain secretes at the top and middle part of the stem. The sclerenchimatic sheath is upper the vascular bundles. Cambium is distinguishable. Secondary thickening develops from a conventional cambial ring. In the lower part of the stem secondary thickening tissues predominate. Vascular bundls consists of secondary phloem and xylem. Xylem has vessels. Vessel end-walls are simple. Vessels have not vestured pits. Pith cells are large and spherical. Cavity arises in the middle and lower part of the stem.

The subterranean rhizome structure is similar to those of the stem. A transverse section taken from the lateral roots was observed (Fig. 3Б). The features of the primary structure are retained. The typical primary root is bounded by an exodermis. The centre of the root made up entirely of many-layered cortex, what have cavities and cells with starch. The inner side is an endodermis. Next follows the pericycle, and then the vascular system is in the middle. Vessel wall pittings and thickenings are similar to those of the stem.

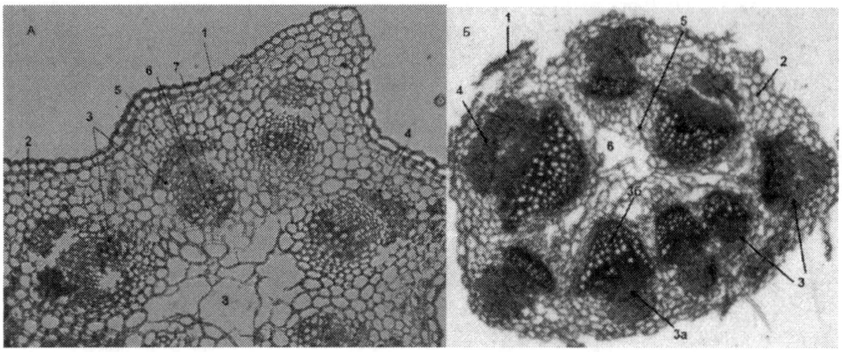

Fig. 2 *Thesium ebracteatum* transverse section of stem A Top part X200 1 – epidermis; 2 – cortex; 3 – vascular bundle; 4 – sclerenchyma; 5 – phloem; 6 – xylem; 7 – cambium; 8 – pith; Б Middle part X100. 1 – epidermis; 2 – cortex; 3 – vascular bundle; 3a – phloem; 3b – xylem; 4 – sclerenchyma; 5 – pith; 6 - cavity.

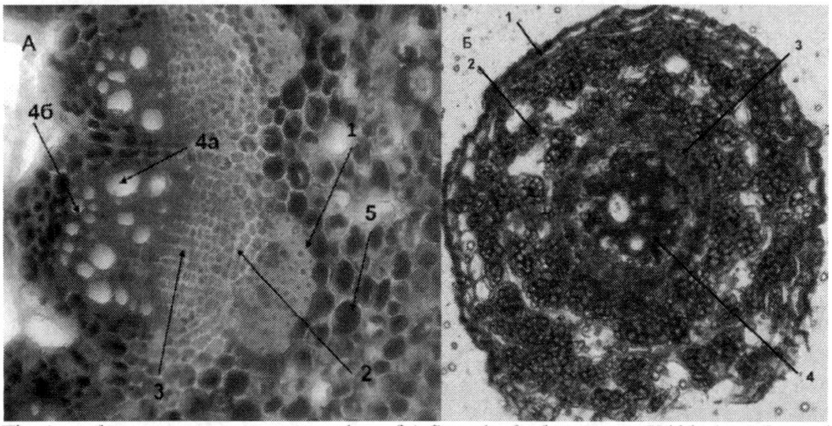

Fig. 3. *Thesium ebracteatum* transverse section of A Stem in the lower part X400. 1 – sclerenchyma; 2 – phloem; 3 – cambium; 4a – secondary xylem; 4б – primary xylem; 5 – starch grains. Б Lateral root X400. 1 – exodermis; 2 – cortex with starch grains; 3 – phloem and pericycle; 4 – xylem.

4 Discussion

Thesium L. with ca. 330 species counts the highest number of species of any genus in *Santalales* [2]. The *Santalaceae* family (*Santalaceae* R.Br.) is represented in Russia by 12 species, 3 species are found in the Urals. All of them are hemiparasitic herbs from the genus *Thesium* L. Several studies demonstrate geographical distribution of *Thesium* species [3-4], biochemical features [5], physiology of host plant-parasitic plant interaction in comparison with others parasitic plants [6], host range and selectivity [7], active ingredients of medicinal material of *Thesium* [8]. This reflects the general trends in the study of the biology of hemiparasitic plants. The analysis of functional features of adaptation to parasitism in plants is considered as importance [9, 11, 14]. The anatomical studies of *Thesium* species in the Urals have never been done before. We performed analysis of available data on the structure of root-hemiparasitic *Scrophs* such as *Rhinanthus*, *Melampyrum*, *Odontites* species and others [10-14] in comparison with *Thesium ebracteatum*. Root-hemiparasitic *Scrophs* have leaves with hydathodes and trichomes [11-13], stems and roots with ring cambium [11, 14] that produce additional xylem tissue embedded lignified xylem parenchyma and deficiently secondary phloem [10]. All roots hemiparasitic plants had haustorial connections, but we noticed that hemiparasitic Scrophs had significant differences from *Thesium* in leaves, roots and stems structure. It is confirmed that not only the parasitic way of life affects the nature of the vegetative organs adaptations. An anatomical structure of leaves, roots and stems reflects primarily belonging to the taxonomic group of hemiparasitic plants.

This study was supported by the Russian Foundation for Basic Research (project no. AAAA-A17-117072810011-1).

References

1. E.J. Salisbury, Philos. Trans. R. Soc. London B, 1-65 (1927)
2. D.L. Nickrent, V. Malécot, R. Vidal-Russell, J. Der, Taxon **59**, 538-558 (2010)
3. A. Romo, Y. Didukh, A. Boratyński, Ann. Bot. Fennici **41**, 273-281 (2004)
4. K.S. Tojibaev, Czech J Genet. and Plant Breeding **46**, 45-46 (2010)
5. X. Zhang, B. Liu, Q. Guo, L. Song, L. Chen, C. Wang, Biochem. Syst. and Ecol. **64**, 46-52 (2016)
6. A. Fer, N. Russo, P. Simier, M.-C. Arnaud, P. Thalouarn, J Plant Phys. **143**, 704-710 (1994)
7. K. Suetsugu, A. Kawakita, M. Kato, Ann. Bot. **102**, 49-55 (2008)
8. F. Luo, Q. Guo, J Zhongguo Zhong Yao Za Zhi. **36**, 2042-6.(2011)
9. J. Těšitel, Plant Ecol. Evol. **149**, 5-20 (2016)
10. O.A. Kiseleva, Bull. Perm Univ. Biol. **3**, 18-26 (2013) [In Russian]
11. J.M. Hibberd, W.D. Jeschke, J Exp. Bot. **52**, 2043-2049 (2001)
12. J.M. Canne–Hilliker, C.M. Kampny, Can. J Bot. **69**, 1935-1950 (1991)
13. R.N. Govier, M.D. Nelson, J.S. Pate, New Phytol. **67**, 963-972 (1968)
14. D.D. Cameron, W.E. Seel, New Phytol. **174**, 412-419 (2007)

BIO Web of Conferences 11, 00023 (2018) https://doi.org/10.1051/bioconf/20181100023
Prospects of Development and Challenges of Modern Botany

Genome composition and assessment of the divergence between Russian boreal species in the genus *Elymus* (Poaceae) detected on the basis of sequencing of the nuclear gene GBSSI

Elena Kobozeva*, *Sergey* Asbaganov, and *Alexander* Agafonov

Central Siberian Botanical Garden SB RAS, Zolotodolinskaya st., 101, Novosibirsk, Russia, 630090

Abstract. A comparative study of the sequence of GBSSI gene fragments in 10 species of the genus *Elymus* growing in the territory of Asian Russia was carried out. Data on the genomic composition and on the levels of interspecific divergence were obtained. This information is a necessary requirement for the construction of phylogenetically oriented taxonomic system of the genus *Elymus* within Russia.

1 Introduction

Serial studies conducted previously confirming that the cytogenetic data for the genomic constitution and evolutionary relationships both among North American [1] and Asian species of the genus *Elymus* [2] significantly complement the molecular phylogenetic analysis of the single-copy GBSSI gene. This paper presents the results of a comparative study of the sequence of the GBSSI gene in 10 *Elymus* species growing on the territory of Siberia and the Russian Far East in order to determine their genomic constitution and to assess the levels of divergence and phylogenetic differentiation. This information is a necessary requirement for the construction of a phylogenetically-oriented taxonomic system of the genus within Russia.

2 Materials and Methods

The plant material of species, which was thoroughly taxonomically identified, was used in the work. The set of samples included species of the genus *Elymus* common in the Asian part of Russia with an unidentified genomic constitution.

All procedures for DNA extraction, cloning and sample preparation for sequencing were performed according to the generally accepted protocols proposed earlier by R. Mason-Gamer [3].

The plasmids were checked for the presence of insertions by electrophoresis in a 1% agarose gel. The products of the Sanger reaction were purified and analyzed on the automatic gene analyzer ABI PRIZM 3130XL on the basis of the Central Committee of the

* Corresponding author: ekobozeva87@mail.ru

SB RAS "Genomics". Chromatograms were subsequently tested and corrected in Unipro UGENE v1.20.0 [4]. From each species sample, 6 clones of the GBSSI gene fragment from 9 to 14 exons were obtained and sequenced [3]. Genetic distances were calculated using the model of the nucleotide substitutions [5]. The construction of dendrograms by the method of Neighbor-Joining was carried out in the program TREECON ver. 1.3b [6]. Gene sequences in species bearing the marker genomes St, H and Y were drawn from the NCBI database.

Accessions and clones of the following species were studied: *E. kamczadalorum* (Nevski) Tzvel. H10432, clones Kamcz_781-786; *E. charkeviczii* Probat. H10470, clones Chark_791-796; *E. kronokensis* (Kom.) Tzvel. KES-9603, clones Kron_801-806; *E. lenensis* (M. Pop.) Tzvel. 12-0125, clones Lenen_811-816; *E. subfibrosus* (Tzvel.) Tzvel., ANA-1181, clones Subfib_821-826; *E. macrourus* (Turcz.) Tzvel. 12-0135, clones Macro_831-836; *E. jacutensis* (Drob.) Tzvel. 13-0443, clones Jacut_841-846; *E. komarovii* (Nevski) Tzvel. GAR-0501, clones Komar_851-856; *E. transbaicalensis* (Nevski) Tzvel. GAR-0530, clones Trans_861-866; *E. sajanensis* (Nevski) Tzvel. ZUN-0502, clones Sajan_871-876.

3 Results and Discussions

The dendrogram constructed by the NJ method (Figure 1) clarified not only the macro-, but also microevolutionary events, confirming our previous results of complex biosystematic studies on the analyzed group of boreal species of Siberia and the Far East of Russia. In all studied species, gene variants were found, unique to St and H genomes. This means that these species belong to the tetraploid StH-genome group. If we examine the location of the samples inside clades of subgenomes, then the clusters combined the species accessions according to the assumed relation. Thus, the species *E. jacutensis* and *E. macrourus* were united in common clusters in both major H- and St-clades, thereby confirming the previously assumptions that *E. jacutensis* is an awned form of *E. macrourus* [7, 8]. This fact agrees with the data on comparative morphological and electrophoretic analysis and hybridization of selective biotypes of these species [9].

Clones of *E. kamczadalorum* were united in a separate cluster, in which the most closely related were the North American species *E. trachycaulus* 3a and 3b, which confirms the hypothesis of a phylogenetic connection between species from Kamchatka and North America. At the same time, the isolation of *E. kamczadalorum* from the Kamchatka species *E. charkeviczii* was confirmed, that was discovered earlier with the help of comparative morphology, endosperm proteins electrophoresis, sexual hybridization [10] and molecular ISSR analysis [11]. Inside the clades of subgenome H, clones *E. charkeviczii* and the reference clone E. mutabilis 2b_H from the Altai Mountains showed a reliable link. This still requires an explanation, taken into account the geographical remoteness of the collection points for biotypes. In the clade of St subgenome, *E. kronokensis*, *E. sajanensis* and *E. lenensis* were united in a common cluster, which also does not contradict our previously obtained data on the phylogenetic proximity of the first two species. Previously, for the new combination *E. lenensis* (M. Pop.) Tzvel. N.N. Tzvelev noted the close relation of this species to *E. kronokensis*. [8].

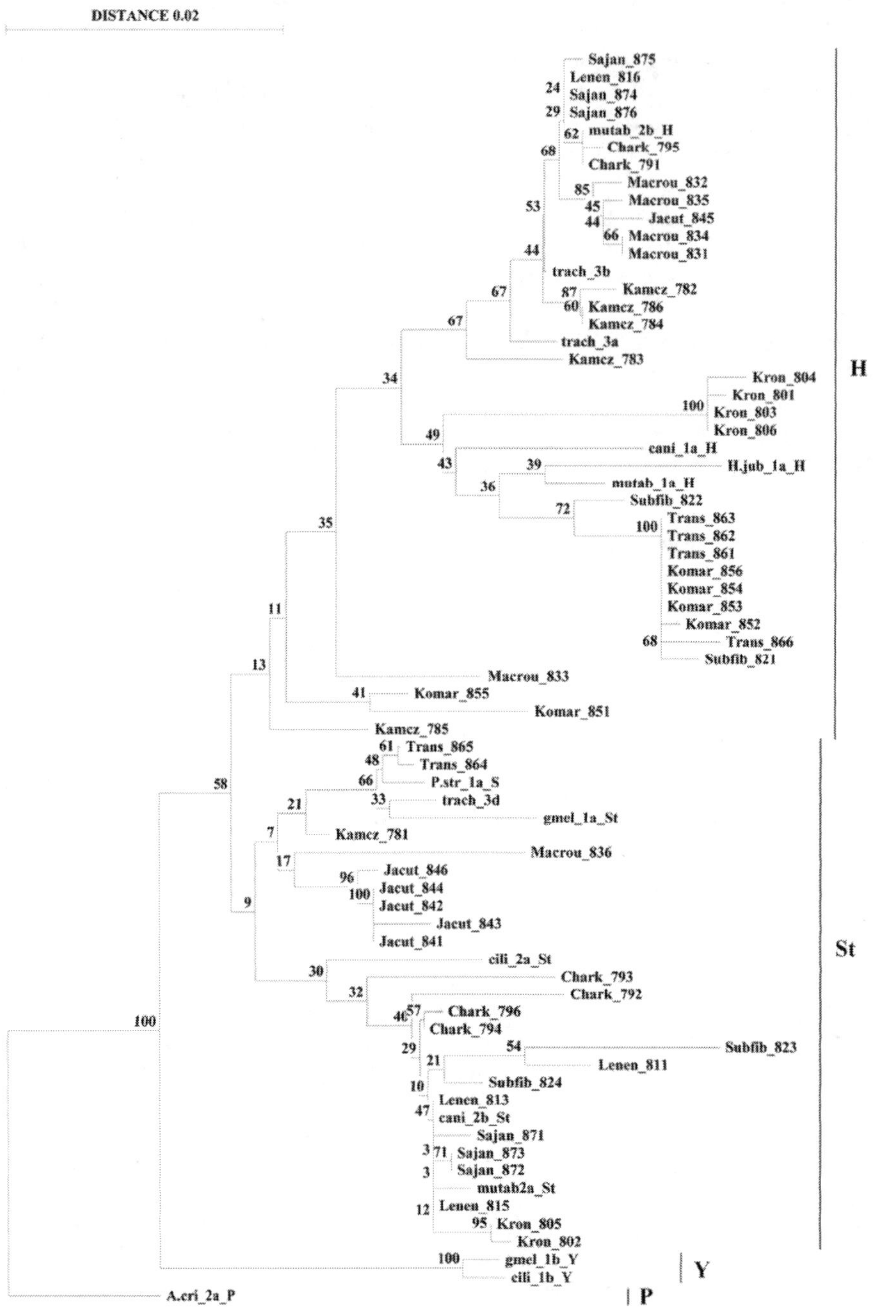

Fig. 1. NJ dendrogram based on the results of sequences of the gene GBSSI exons in 10 species of the genus *Elymus* from the Asian part of Russia. Clones drawn from the NCBI: *E. trachycaulus* (trach_3a_H, trach_3b_H, trach_1b_S); *E. caninus* (cani_2b_St, cani_1a_H); *E. mutabilis* (mutab_2a_H, mutab_2b_H); *E. ciliaris* (cili_2a_St, cili_1b_Y); *E. gmelinii* (gmel_1a_St, gmel_1b_Y); *Hordeum jubatum* (H.jub_1a_H); *Pseudoroegneria strigosa* (P.str_1a_S); *Agropyrum cristatum* (A.cri_2a_P).

Species *E. komarovii* and *E. transbaicalensis*, as expected, formed a common cluster. One of the reasons was the joint growth of these biotypes in relatively close populations.

Since in mixed populations of these species there is active hybridization and introgression, the results of the analysis require confirmation with the use of accessions from different locations of growth of these species, both jointly and separately. According to our long-term observations, the species *E. komarovii* and *E. transbaicalensis* are genetically independent, but not always morphologically distinct. Modern processes of hybridization and introgression can lead not only to new interspecific biotypes and micropopulations, but also to the formation of new species.

One of the most difficult species for identification is *E. subfibrosus*. The accession collected in the regional population of Anadyr (Chukotka) is grouped in a clade of H-genomic clones together with morphologically close *E. komarovii* and *E. transbaicalensis* from Siberia. Obviously, for study of such polymorphic species with a wide area in Siberia and the Russian Far East more biotypes from different populations are needed.

To construct a new phylogenetically oriented genus system, as the first step, it is necessary to classify intra-genus taxa based on the genomic constitution of the species. The generic system should be based on the taxonomic model of microevolutionary groups, which are evolving through hybridization and introgression. The degree of relationships between taxa can be determined using biosystematics methods with the determination of the crossability of individuals. Taking into account previous biosystematic studies, we proposed the principle of constructing a model of microevolutionary groups of some boreal allotetraploid *Elymus* taxa in the territory of Russia. Each microevolutionary group should be considered as a branched system of species and intraspecies taxa. Therefore, the system of the genus will be presented as phylogenetically confirmed structure.

Taxonomic ranks of different microevolutionary groups could be defined as subspecies, sections, subsections or aggregate species in a broad sense. Probably, it might be appropriate to recognize subgenera within the genus *Elymus*, since the species growing on the territory of Russia according to recent treatments [12] were combined into sections without consideration of their genomic constitution.

References

1. R.J. Mason-Gamer, Syst. Bot. **26**, 757–768 (2001)
2. R.J. Mason-Gamer, M. M. Burns, M. Naum, *Mol.* Phylogenet. Evol. **54**, 10–22 (2010)
3. R. J. Mason-Gamer, C.F. Weil, E.A. Kellogg, Mol. Biol. Evol. **15**, 1658–1673 (1998)
4. K. Okonechnikov, O. Golosova, M. Fursov, Bioinformatics **28**, 1166–1167 (2012)
5. M. Kimura, J. Mol. Evol. **16**, 111-120 (1980)
6. Y. Van de Peer, R. De Wachter, Computer Applications in Biosciences **10**, 569–570 (1994)
7. N.N. Tzvelev, Arkticheskaya flora SSSR **2**, 230–247 (1964) (in Russian)
8. N.N. Tzvelev, Novosti sistematiki vysshikh rastenij **10**, 19–59 (1973) (in Russian)
9. A.V. Agafonov, Plant Life of Asian Russia **2**, 20–32 (2008) (in Russian)
10. A.V. Agafonov, D.E. Gerus, B. Salomon, Czech J. Genet. Plant Breed **41**, 79–85 (2005)
11. E.V. Kobozeva, M.V. Emtseva, S.V. Asbaganov, A.V. Agafonov, Plant Life of Asian Russia **3**, 43–50 (2017) (in Russian)
12. N.N. Tzvelev, N.S. Probatova, Komarovskiye chteniya **57**, 5–102 (2010) (in Russian)

Assessing species resistance and productivity of the genus *Paeonia* L. under adaptation to continental climate conditions

Olga Komina*

Central Siberian Botanical Garden of the Siberian Branch of the Russian Academy of Sciences, Novosibirsk, Russia

Abstract. The paper presents results of multiyear work to create and study the collection of *Paeonia* L. species in the Central Siberian Botanical Garden of the Siberian Branch of the Russian Academy of Sciences (CSBG SB RAS) (Novosibirsk), which numbers 13 species of 4 botanical sections nowadays. All herbaceous species of the genus *Paeonia*, both geophytes and hemicryptophytes, have been safely wintered during 12 years without additional shelter. The study shows assessment results of prospects of the genus *Paeonia* species representatives. It gives recommendations on landscape design for *Paeonia* species selection.

1 Introduction

The genus *Paeonia* L. is considered a leading floral culture, it is an inalienable favorite in spring garden [1]. Above 1,000 varieties are grown in Russia today, but recently, due to the changing trends in landscape design, wild *Paeonia* representatives have become very popular.

The selection criteria of introduction deserving plants are always associated with the evidence of possibilities to grow and breed under new growing conditions, which in turn, is determined by the introducent ability to adapt to other climatic conditions, and resist various pathogens. When working with ornamental plants, the selection criteria is not only an original color, exquisite shape and large dimensions of flowers and inflorescences, but economically valuable qualities, such as the rate of growth, flowering abundance and duration, decorative habitus.

The study objective is to assess the adaptive potential of the genus *Paeonia* L. species under continental climate conditions of West Siberian forest-steppe.

2 Material and methods

Observations and experiments have been carried out since 2007 based on the collection of the *Paeonia* generic complex of CSBG SB RAS, the collection of living plants in open and covered ground (No USU 440534). The research is gone on the project "Identifying the

* Corresponding author: kuznetsova_olga8@mail.ru

ways of plant adaptation to contrasting habitat conditions at population and organism levels" AAAA-A17-117012610053-9 (State Registration No).

All plants have been gathered in natural habitats. The assessment of prospects to introduce peony species of native and non-regional flora has been done in accordance with R.A. Karpisonova's methodology [2].

3 Results

At present gathered and studied *Paeonia* collection of CSBG SB RAS has 13 species from 4 botanical sections, as well as 56 varieties of 5 garden groups (species *P. biebersteiniana* Rupr., *P. mollis* Anderson. are at the primary introduction stage).

While introducing economically valuable plants under continental climate conditions of West Siberian forest-steppe, the main selection criterion is winter resistance, which refers to introducent ability to withstand complex environmental factor impacts in pre-winter, winter and early spring periods.

The genus *Paeonia* most herbaceous representatives are geophytes, their buds of renewal are at 5 cm depth of soil surface, that creates favorable conditions for wintering. Some species are hemicryptophytes, e.g. *P. obovata* Maxim. growing at the Far East under monsoon climate conditions.

During 12 years of studying all herbaceous species of the genus *Paeonia* safely wintered without additional shelter. The main problems have appeared under attempts to introduce dendriform species of *Paeonia* (*P. suffruticosa* Andr.), which need to save an aboveground shoot system during wintering season for adequate flowering the next year.

To select the *Paeonia* species material for introducing in CSBG SB RAS, the laboratory is guided by the following principles in forming collections of ornamental plants of natural flora [3]: making collections of ornamental plants of the natural flora in frames of generic complexes or their fragments, it is necessary to represent various pheno-rhythm types, life forms and ecological groups. The studied species grew under conditions of temperate continental climate in steppe and mountainous regions, as well as in the temperate monsoon climate (a forest belt). As our introductory experiments show, two pheno-rhythm types were distinguished in our objects during adaptation to the seasonal specifics of natural habitats humidification conditions: spring-summer green and spring-summer-autumn green. However, *P. tenuifolia* L. vegetated as a hemiephemeroid under moisture deficit in late autumn and early summer some years.

Criteria for selecting *Paeonia* breeds for the regular style of decoration used in greening the modern urban environment differ greatly of the landscape style requirements [4], which becomes more popular among landscape designers. Related to this we have described biological features of *Paeonia* species, which create both an early summer aspect of flowering, and the decorative foliaceous effect. After a long winter season, the human visual perception in an urbanized environment is searching for early spring flowers and growing shoots. Representatives of the genus *Paeonia* belong to the number of not only early growing ornamental perennials. Most *Paeonia* species, in contrast to breeds, unfurl leaves of original shape and color almost a few days after regrowth.

The final introduction success is evaluated by a set of signs, the most important of which are: completeness of passing a big (ontogenetic) and little life cycle by plants, inherent habitus preservation. When evaluating the *Paeonia* species of local and non-local flora, each feature was assessed on a three-point scale. As Table 1 shows, complete generative development (which determines seed reproduction) was absent in such species as *P. mlokosewitschii* Lomak, *P. wittmanniana* Hartwiss ex Lindl., *P. lutea* Delavay ex Franch (refers to tree-like plants). In different years these species either did not bloom, or bloomed without forming full seeds. Seed production with a low number of seeds was not observed annually in such species as *P. tenuifolia*, *P. obovata*, *P. oreogeton* S. Moore. But *P. lactiflora* Pall., *P. anomala* L., and *P. hybrida* Pall. had abundant and annual seed production.

Table 1. Assessment of species resistance and productivity of the genus *Paeonia* of local and non-regional flora by R.A. Karpisonova's methodology

I.Generative development determining seed reproduction:	1 – seed production is absent (plants do not bloom; bloom, but seeds do not form; the seeds do not ripen) – *P. mlokosewitschii* – *P. wittmanniana* – *P. lutea* (refers to dendritic)	2 – seed production is not annual, few seeds – *P. tenuifolia* – *P. obovata* – *P. oreogeton*	3 – abundant and annual seed production – *P. lactiflora* – *P. anomala* – *P. hybrida*
II.Vegetative reproduction:	1 – is absent (plant paresis is not observed) – *P. lactiflora* – *P. mlokosewitschii* – *P. oreogeton* – *P. wittmanniana* – *P. obovata* – *P. anomala* – *P. hybrida* – *P. tenuifolia* – *P. lutea*	——	——
III. Preserving habitus and life-forms in culture:	1 – plants are low, do not reach their size in nature – *P. lutea*	2 – retain natural sizes – *P. tenuifolia* – *P. obovata* – *P. oreogeton* – *P. wittmanniana* – *P. mlokosewitschii*	3 – exceed – *P. lactiflora* – *P. anomala* – *P. hybrida*
IV.Plants survival of in unfavorable seasons (determined by recalculation):	1 – considerable annual dying off shoots and specimens ——	2 – shoots and specimens die only in especially severe winter – *P. mlokosewitschii* – *P. wittmanniana* – *P. lutea*	3 – plants do not drop off – *P. lactiflora* – *P. tenuifolia* – *P. anomala* – *P. hybrida* – *P. oreogeton* – *P. obovata*

Natural vegetative reproduction (plant paresis) was not observed in *P. lactiflora*, *P. mlokosewitschii*, *P. oreogeton*, *P. wittmanniana*, *P. obovata*, *P. anomala*, *P. hybrida*, *P. tenuifolia*, *P. lutea*. To accelerate *Paeonia* species reproduction aimed their conservation *ex situ*, we made artificial vegetative propagation by a bush dividing.

When evaluating the habitus and life form preservation in culture, *P. lutea* got 1 point, as it did not develop a heavy system of aboveground skeletal axes, and did not reach sizes inherited in nature.

P. tenuifolia, *P. obovata*, *P. wittmanniana*, *P. mlokosewitschii* retained their natural sizes (2 points). Only *P. lactiflora* and *P. anomala* exceeded their inherent natural size (maximum score of 3 points), which might be due to similarity of natural habitat conditions against and competition lack in a cenosis, as well as a high level of agricultural technology in cultivation.

Practically, all studied species were character a high survival rate of plants in unfavorable seasons. None of 9 species was marked by annual specimen dying off. Individual shoots dying off was observed in *P. mlokosewitschii*, *P. wittmanniana*, and *P. lutea* only in extremely harsh winters with a sharp long temperature decrease in November and insufficient snow cover; they received 2 points according to this sign. *P. lactiflora*, *P.*

tenuifolia, *P. anomala*, *P. oreogeton*, *P. obovata* obtained the maximum score (3 points), as they did not show dropping off shoots or specimens.

Cluster analysis showed that the studied species-introducents of the genus *Paeonia* should be shared three groups according to the degree of their outlooks for West Siberian forest-steppe.

The first group includes 3 species *P. lactiflora*, *P. anomala*, P. *hybrida* – received the greatest assessment of prospects according to R.A. Karpisonova's scale (10 points). All of them are marked as species growing under the natural conditions in Siberia [5]. This is a local species of *P. anomala*, as well as *P. hybrida* growing close to Novosibirsk Region (Altai Republic, Altai Territory), and *P. lactiflora*, a species introduced from Primorye with a wide range (East Siberia-Far East).

The second group consists of two Primorye species, *P. oreogeton*, *P. obovata* growing in nature at the Far East, and *P. tenuifolia* from the Stavropol Region, these 3 species are confined to a fairly wide range of habitats, which indicate their broad ecological plasticity (8 points each).

The third group includes two Caucasian species, *P. wittmanniana* and *P. mlokosewitschii*, which natural habitat conditions differ greatly from the climatic conditions of West Siberian forest steppe (*P. mlokosewitschii* is endemic of Central and Eastern Caucasus) (6 points each). *P. lutea* enter this same group as well; it received the smallest score (5 points) on stability and productivity in our region, that is not surprising for a species related to tree-like plants, which do not retain their biomorphs inherited in nature under severe climatic conditions of West Siberian forest-steppe.

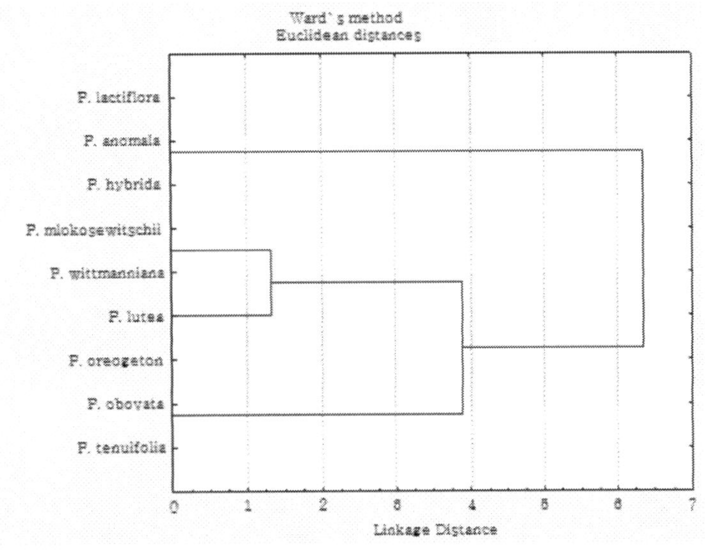

Fig. 1. Cluster analysis on resistance and productivity parameters of the genus *Paeonia* species of local and non-regional flora.

4 Discussion

The comparative assessment of prospects revealed that a local species *P. anomala*, as well as the tested non-regional species (*P. lactiflora*, *P. mlokosewitschii*, *P. oreogeton*, *P. wittmanniana*, *P. obovata*, *P. hybrida*, *P. tenuifolia*, *P. lutea*) showed themselves highly plastic adapted to climatic conditions, which can successfully grow in West Siberian forest-steppe. *P. oreogeton*, *P. mlokosewitschii* and *P. wittmanniana* need preliminary choice of optimal microecological conditions, after which they bloom successfully and bear seed regularly.

P. tenuifolia, which does not form self-sowing (by the way, it is not a positive sign for flower gardens), does not concede to species forming high, albeit longer, flowering effect.

In analyzing the genus *Paeonia* species stability and productivity compared with similar studies by other researchers, it was concluded that *P. anomala* is the most deserving plant for both Bashkir Pre-Ural forest-steppe [6, 7], and Belarus' [8], in Kemerovo Region it is even recommended for plantation cultivation [9].

To study plant biodiversity conservation *ex situ*, the methodology by R. A. Karpisonova [1] is the most informative to assess the adaptive features of perennial grass polycarpic species. Evaluating the prospects of *Paeonia* species as ornamental plants it is necessary to pay attention to such decorative elements as leaf and fruit shape, their color in addition to the flowering effect.

5 Conclusion

Thus, while determining resistance and productivity, *P. lactiflora*, *P. anomala*, *P. hybrida* have got the highest scores (10 points); they exceed the inherited bush size and maintain abundant and annual seeding.

P. tenuifolia, *P. obovata*, *P. oreogeton* (8 points) are considered promising as well, because they retain the bush natural size and form seeds being introduced into more severe climatic conditions.

Such *Paeonia* species as *P. lactiflora*, *P. anomala*, *P. hybrida* (10 points), *P. tenuifolia*, *P. oreogeton*, *P. obovata* (8 points) occur highly winter-resistant; *P. mlokosewitschii*, *P. wittmanniana* (6 points), and *P. lutea* (5 points) are rather winter-resistant (particular units drop off). Cluster analysis has confirmed the obtained experimental data.

None of 9 species does reveal specimen dying off. In general, the *Paeonia* species should be regarded sufficiently winter-resistant objects, which give a high decorative effect not only in the flowering period, but as well due to a heavy compact bush habitus and beautiful foliage. However, their decorative effect can reduce by diseases and pests under local conditions.

For *Paeonia* species identified as promising it is useful to recommend the most effective combinations with other less common ornamental plants. For instance, mixborder bright fragments will be small groups formed by *P. tenuifolia*, *Trollius asiaticus* L., *T. europaeus* L.; *P. obovata*, *Primula juliae* Kusn., *Primula pallasii* Lehm.; *P. lactiflora*, and *T. ledebourii* Reichb.

This assortment of peony species may be successful in applying to create wildlife fragments in the urbanized urban environment, which are popular landscape design direction nowadays.

References

1. R. Kamenetsky, J. Dole, Floriculture Ornament. Biotechnol. **1**, 6 (2012)
2. R.A. Karpisonova, *Herbaceous plants deciduous forests of the USSR: ecologic-floristic and introduction characteristics* (Nauka, Moscow, 1985) in Russ.
3. T.I. Fomina *Biological characteristics of ornamental plants of natural flora in West Siberia* (Geo, Novosibirsk, 2012) in Russ.
4. N. Müller, H. Sukopp, Landscape Ecol. Eng. **12**, 1 (2016)

5. N.V. Friesen, *Paeonia* L. – Peony, *Flora Sibiriae*. (**6**, Nauka, Novosibirsk, 1993) in Russ.
6. A.A. Reut, Bull. OGU. **12** (2017) in Russ.
7. A.A. Reut, L.N. Mironova, Ital. Sci. Rev. **7**, 16 (2014)
8. V.V. Gayshun, Proc. Natl Acad. Sci. Belarus. **4** (2015) in Russ.
9. Y. V. Zagurskaya, I. N. Yegorova, Bull. ASAU. **12**, (2015) in Russ.

BIO Web of Conferences **11**, 00025 (2018) https://doi.org/10.1051/bioconf/20181100025
Prospects of Development and Challenges of Modern Botany

Drought resistance of *Astragalus cicer* and *A. sulcatus* at the initial stages of plants development

Tatyana Kornievskaya[1,*]

[1]Altai State University 656049, Barnaul, Lenina Str., 61, Russia

Abstract. The article considers laboratory evaluation methods of water stress influence on the initial stages of Astragalus cicer and A. sulcatus development by seeds ability to germinate in osmotic sucrose solutions, characteristics of radicles and depression of shoots growth processes induced by the solution osmotic pressure. Osmoregulatory characteristics of seeds are estimated. It was ascertained that the rise in solution osmotic pressure up to 2 atm leads to a twofold and fivefold decrease in seeds germinating capacity in A. cicer and A. sulcatus respectively. At the sucrose concentration level of 4 atm, seeds do not germinate at all. Strong concentrations of osmotic solutions cause the depression of shoots growth processes and the reduction of astragaluses dry biomass.

1 Introduction

Territories belonging to various soil-geographical regions of our planet are constantly or periodically exposed to drought, which leads to significant crop losses. The total area of the arid territories having less than 250 mm annual precipitation constitutes more than a third of the earth's surface. First of all, drought deteriorates the seeds germination and the young crops condition, as long as the initial stages of plants development are the most vulnerable. Young shoots lose the tie with the seed and become completely dependent on the external factors, i.e. mineral and water supply, air and soil humidity. Drought causes stomata of plants to close, which in its turn leads to the decline in the photosynthetic efficiency, productivity and crops quality [1].

The reaction of different plant species to the lack of water depends on the level of their sustainability, which quantitative measure is the degree of productivity decrease in response to the stress pressure [2]. Drought resistance is an important ecological adaptation determining the combination of genes that influence biochemical, physiological and morphological characteristics of the individual.

Drought resistance evaluation of the plant species and cultivars with the application of indirect laboratory methods is becoming the most popular nowadays.

Laboratory methods allow detecting the seeds and shoots condition by means of their cultivation in osmotic sucrose solutions imitating water deficiency. These methods allow us

* Corresponding author: galtsovatw@yandex.ru

to estimate the plants ability to withstand the lack of moisture at the early stages of ontogeny within a short period regardless of the weather conditions. S. Grzesiak [3] consider that laboratory methods of plants drought resistance evaluation with the use of osmotic solutions (sucrose, mannitol, etc.) are an adequate criterion to determine plants sustainability to water deficiency and are comparable to direct methods.

Several scientists [4] noted the dependence of plants drought resistance on the complex of morphological and physiological parameters. The rate of root system growth and vigor as special means of obtaining water and mineral substances are regarded as the most significant traits.

A. Bukhinger was the first to suggest the idea to evaluate the drought resistance of plants by the suction force that germinating seeds develop in the sucrose solution. The higher the root system growth rate and the greater the rooting depth are, the more the drought resistance is [5].

2 Materials and methods

The material of the study was the seeds of two astragalus species being introduced in the dry steppe conditions of Mikhaylovskiy region (Altai krai) – *Astragalus cicer* L. and *A. sulcatus* L. gathered in 2017. The evaluation of water stress influence on the growth and development of astragaluses was carried out by means of three indirect methods: seeds ability to germinate in osmotic sucrose solutions, characteristics of radicles and depression of shoots growth processes caused by the osmotic pressure of the solution [6].

For the purpose of the evaluation of astragalus drought resistance, diluted concentrations of sucrose solutions with the osmotic pressure of 2 and 4 atmospheres were selected. The sucrose mass fraction in the solution was determined with the use of the van't Hoff equation in accordance with the target osmotic pressure by the formula:

$$P_{osm} = C_{м} \times R \times T\,(Pa) \qquad (1)$$

P_{osm} – osmotic pressure (Pa);
$C_{м}$ – sucrose molality (mol/L);
R – gas constant (J/mol × K);
T – temperature (Kelvin), at +20°C.

The viability of seeds was estimated by the Yu.D. Akhlamova procedure [7].

The seeds, preliminarily scarified by concentrated sulphuric acid, were placed on filter paper in Petrie dishes (50 seeds per dish) with four-fold replication (for the osmotic sucrose solutions) and two-fold replication for the reference sample (distilled water). The seeds were poured over up to maximum water holding capacity with 5 ml of the sucrose solution with the respective osmotic pressure (experiment) or water (reference).

The seed sprouting was carried out at room temperature of +19–20°C. Germinative capacity was determined on the seventh day.

In order to predict the drought resistance of plants in early ontogeny in the conditions of insufficient moisture supply, we studied the peculiarities of the root biomass formation. On the seventh day, we counted the number of the radicles (pcs), measured the radicles mass (mg) and their length (mm). Different characteristics of the radicles in the sucrose solutions were compared to the reference sample (water).

After determination of the seed germinative capacity, the radicles and shoots in each Petrie dish were cut down, put into weighing bottles, placed into an oven and dried for 3 hours at the temperature of +105°C. After drying, the material was cooled in the desiccator and weighed on the analytical balance. The average dry mass of the sprouts per one replication in the reference sample was set at 100% (x), while the average dry mass in the experiment (y) was determined as percentage of the dry mass in the reference sample.

The depression degree (Z) in the accumulation of dry mass by the shoots at high osmotic pressure was calculated by the formula:

$$Z = 100 - y/x \times 100 \tag{2}$$

On the basis of three considered methods, we differentiated between four groups of shoots resistance: I – drought resistant (80–100%), II – resistant (60–79%), III – low resistant (40–59%), IV – non-resistant (0–39%).

3 Results of the research

Based on the laboratory germinative capacity of the astragalus seeds in concentrated sucrose solutions, we noticed a tendency towards its decrease if the concentration of the solution increases. The seeds of *Astragalus cicer* and *A. sulcatus* in the reference sample were characterized by high percentage of germinative capacity: 88% and 76% respectively. Increase in the concentration of the solution up to 2 atm leads to a significant decline in germinative capacity of *Astragalus cicer* (two times less – 42%) and to a five-time fall in germinative capacity of *Astragalus sulcatus* (15%) (Table 1). At the sucrose concentration level of 4 atm, seeds do not germinate at all. This indicates the low sustainability of seeds and shoots to water deficiency.

Table 1. Influence of the osmotic pressure on the formation of radicles and aboveground sprout shoots of *Astragalus* species.

Species	V, %	Germinative capacity, %			Roots length, mm			Sprouts length, mm		
		RS	P_{osm}=2 atm.	P_{osm}=4 atm.	RS X ± m	P_{osm}=2 atm. X ± m	P_{osm} =4 atm.	RS X ± m	P_{osm}=2 atm. X ± m	P_{osm} =4 atm.
A. cicer	97	88	42	0,25	7,6±0,38	4,5±0,33	2,0	10,5±0,42	5,97±0,22	4,0
A. sulcatus	96	76	15	0,50	4,0±0,45	2,2±0,18	1,0	7,8±0,34	4,6±0,39	2,0

Note to Figure 1: V – viability, RS – reference sample (distilled water), P_{osm} – osmotic pressure of the sucrose solution.

Water stress causes the growth inhibition of the aboveground and underground plant organs, which is reflected in the overall condition of the shoots. The examined astragaluses samples were characterized by the depression of growth processes in the solutions with elevated concentration of sucrose. We observed a twofold decrease in size and biomass of shoots and radicles in comparison with the reference sample if the osmotic pressure was strong (Table 1, 2).

Table 2. Influence of water stress caused by high osmotic pressure of the sucrose solution on the growth processes depression in astragaluses.

Species	Absolute dry roots biomass, mg			Absolute dry sprouts biomass, mg			Degree of growth processes depression (Z), %	
	Reference sample	P_{osm}=2 atm.	P_{osm}=4 atm.	Reference sample	P_{osm}=2 atm.	P_{osm}=4 atm.	roots	sprouts
A. cicer	14,25	1,17	–	117,00	43,90	–	91,79	62,48
A. sulcatus	53,43	1,40	–	30,41	6,50	–	97,38	78,63

The determination of the drought resistance by the depression of the growth processes in plants at the initial stages of ontogeny depends on the rate of the dry mass accumulation by shoots in the conditions of insufficient water supply. Imitated drought influenced the accumulation of biomass in both aboveground and underground parts of the shoots. In case of water deficiency, the root system of astragaluses demonstrated the highest degree of growth processes depression (91,79% – *A. cicer*, 97,38% – *A. sulcatus*), which speaks in favor of a greater degree of the growth processes inhibition in roots as compared to the aboveground vegetative plant parts. A high depression level in shoots is indicative of astragaluses susceptibility to water deficiency.

4 Conclusions

We determined the astragaluses seeds ability to germinate in osmotic sucrose solutions. The osmotic pressure of 4 atm leads to a complete suppression of the growth processes and inability of the seeds to germinate, which indicates low osmoregulatory characteristics of the latter.

The size and biomass of the shoots are significantly smaller in the conditions of water deficiency. Depression of the growth processes in radicles demonstrated a low water potential of the shoots in case of elevated concentration of water solution. The astragalus shoots have low suction force, which manifests their low drought resistance.

The considered complex of laboratory methods evaluating the astragaluses drought resistance at the initial stages of ontogeny allows us to refer *A. cicer* and *A. sulcatus* to the group of non-resistant species.

References

1. S.A. Anjum, X. Xie, L. Wang, M.F. Saleem, C. Man, Afr. J. of Agric. R. **6**, 2026-2032 (2011)

2. G.V. Udovenko, Proceed. on appl. bot., gen. and breed **64**, 5-22 (1979) (in Russian)

3. S. Grzesiak, G. Skrudlik, W. Filek, B. Niziol, J. Agr. C. Scien. **177**, 245-252 (2008)

4. G.G. Borisova, M.G. Maleva, N.V. Chukina, *Plant and stress: course of lectures* (Ekaterinburg, 2008) (in Russian)

5. E.V. Ionova, E.I. Nekrasova, Grain Econom. Rus. **5**, 12-21 (2013) (in Russian)

6. O.V. Bychkova, E.A. Sharlaeva, *Practical course in ecology. Evaluation of plants sustainability* (Barnaul, Altai State University, 2015) (in Russian)

7. Yu.D. Akhlamov, *Patent of Russian Federation 2477595* (2013) (in Russian).

Seasonal rhythm of development of *Allium trautvetterianum* in the Kulyab Botanical Garden

Shomurod Kullaev[1*], *Vera* Cheryomushkina[2†] , and *Mariyo* Boboev[3 ‡]

[1]Rudaki Kulyab State University, 735360 Kulyab, Tajikistan
[2]Central Siberian Botanical Garden SB RAS, 630090 Novosibirsk, Russia
[3]Khatlon Science centre AS RT 735360 Kulyab, Tajikistan

Abstract. The article deals with the seasonal rhythm of development of the bulbaceous acrosympodially accreting polycarpic *Allium trautvetterianum* Regel (the subgenus *Melanocrommyum*) when growing in the Kulyab Botanical Garden of the Khatlon Scientific Center, Academy of Sciences of Tajikistan in 2010-2015. Duration of vegetation and phenological phases is described. The species is characterized on the phenorhytmotype as a shortgrowing ephemeroid. Influence of weather conditions on duration of phenological pahses is shown. It is established that the rhythm of development completely corresponds to the eastern variant of Mediterranean climate.

1 Introduction

Seasonal rhythm of development of plants is a representation of their life cycle which is shown in repetitive phenophases. Study of phenophases is necessary for understanding for plant community functioning [1]. The change in time and duration of the passage of phenophases to adjust to new conditions is one of the indicators of plant adaptation to changing environmental conditions. In the context of climate change, the importance of phenological study is currently increasing.

According to current data the genus *Allium* L. numbers about 900 species. Onions are widespread in the northern hemisphere and represent one of the largestgenera of the world flora [1]. The subgenus *Melanocrommyum* contains 170 species, the range of which is mainly restricted to the region of Ancient Meditarrenean [2]. Thirty one species of this subgenus grow in Tajikistan.

It is known that *Melanocrommyum* species on the rythm of development belong to ephemeroids. A short period of vegetation and duration of stay in dormancy as a bulb most of the year are typical of them [3, 4]. Information on seasonal development of some onions of the subgenus *Melanocrommyum* is mainly availablein the works on introduction of them in different regions of Russia [4-6]; Uzbekistan [3]; the Ukraine [7]; and Germany [8].

* Corresponding author: kullaev_79sh@mail.ru
† Corresponding author: cher.51@mail.ru
‡ Corresponding author: b_mario@mail.ru

Although *A. trautvetterianum* Regel. has long been in use as a food and ornamental plant, its seasonal development pattern is not practically studied.

The aim of the work was study of the seasonal rhythm of development of *A. trautvetterianum* development in the Kulyab Botanical Garden (Southern Tajikistan).

2 Material and methods

Allium trautvetterianum is an endemic plant of the south-western Pamir-Alay, it grows in the semisavanna belt with domination of *Prangos pobularia, Organum tyttanthum, Salvia sclarea, Hypericum scabrum, Taeniatherum asperum, Poa bulbosa,* and *Carex pachystylis,* in the shibliak belt with domination of *Pistacia vera, Crataegus pontica,* and *Amygdalus bucharica,* and occurs at the exposure of vaiegated at an altitude of 700–1700 m above sea level.

The species has been cultivated in the Kulyab Botanical Garden since 1999 г. Onions were collected on the Dzhilantau Ridge (Temurmaliksky Region, Southern Tajikistan) [9]. Development of *A. Trautvetterianum* was observed durring 5 years from 2010 to 2015. Seasonal development was studied by the method of I.N. Beideman [10]. Duration of a concrete given phase was determined by development of 10–20 individuals per introduction population. Belonging to the phenorhythmotype was determined according to the I.V. Borisova classification [11].

The following phases of phenological development: vegetation, duration of functioning of the assimilation apparatus, growth of the flower stalk, budding, blooming, fruiting and dissemination were identified because of development pattern of onions as rosette plants .

Annual meteorological data of the research location were taken from the summaries of Kulyab weather station.

3 Results and discussion

I.G. Serebryakov [12] considered a seasonal rhythm of plant development to be one of the features reflecting adaptation to ecological-phytocoenotic and climatic conditions of growth.

The climate of Tajikistan generally belongs to the eastern variant of the Mediterrenean climate [13]. Kulyab and Kulyab Botanical Garden are located in southern Tajikistan at an altitude of 640 m above sea level in the valley of the Yakhsu River. Kulyab is characterized by insufficiantly humid climate, dry summer, moderately mild winter. The mid-annual air temperature is 16,4°C, sometimes in summer it rises up to 44–46°C, in winter, in certain years, it falls to 17–21°C. Duration of the frost-free period is 260-309 days, and the sum amounts to 5800–6000°. The period with temperature of more than 10°C lasts 240–250 days. The effective heat sum for this period is 2800–3000°. In the hot season often blow dry southern winds in the form of dust storms. Large temperature changes are observed not only during a year, but also during a day. There is a lot of precipitation– 529 mm, but its distribution is very uneven. Up to 70 % of the annuak norm of precipitation falls in winter and spring [14]. Very little rain falls in summer. The period from June to September, at high air temperature, it is almost or rainless at all. The soil is dried out. Stable snow cover, only some centimeters, is formed very seldom.

Allium trautvetterianum is a perennial herbaceous, acrosympodially accreting, nonparticulating bulbous polycarpic plant. Annually unfolds one rosetted shoot consisting of a shortened part with 3 scale-shaped and 2-3 assimilative leaves and leafless flower stalk. The length of a flower stalk reaches 50 cm. Leaves are lanceolate, upward, 2–3cm wide, with an axil up to 3-6 cm. A basal part of the shoot forms a tunicate ovate bulb 1,5-2,5cm in diameter. The bulb is formed by membranous cover scale, storing scale, vaginate scale and axils of green leaves. Outside it is covered with greyish, almost papery splitting sheaths – leaf remnants of previous shoots. Additional roots are thin numerous, superficially arranged, develop on the bulb stem and are annually substituted by new ones. An inflorescence is dense many-flowered globe-shaped or umbellate thyrse. Like most species of the subgenus *Melanocrommyum,* it belongs to ephemeroids by the seasonal rhythm of development. A regeneration bud is laid in the bulb 3 years before development. Generative organs are completely formed by autumn previous to the growing season.

For 6 years of observations vegetation of *A. trautvetterianum* lasted from 71 to 81 days in the Kulyab Botanical Garden. However, vegetation periods differed substantially in various years. The earliest plant growing and emergence of the first leaves were noted in the first decade of February in 2010 and 2015, and the latest one – in the first decade of March in 2013–2014 (Fig. 1).

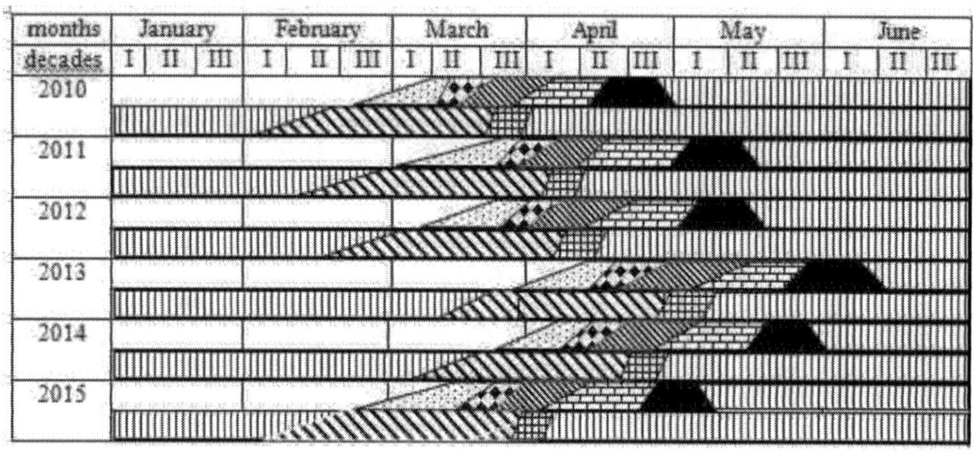

Fig.1. Phenological rhythm of Allium trautvetterianum Regel development in cultivation in 2010-2015.

Legend – – – dormancy, – vegetation, – flower stalk growth, – budding, – blooming, – fruiting, – dissemination, –drying up of leaves.

Such a delay was linked with late onset of spring in 2013–2014 because of negative temperatures in February. This led to the shift of all phases of seasonal development on the later time in those years. In typical years the flower stalk of the generative shoot emerges 2–2.5 weeks after the beginning of vegetation (late February– early March). It increases in length throughout 15–20 days. A budding phase of a certain plant is short and does not exceed 5 days. Plants begin to bloom in mid-March at daily average temperature +10°C. Air temperature affects the beginning of blooming. So, the low daily average temperature (+4,8°C) in March 2012 delayed transition of plants to flowering almost for 10 days. Duration of blooming of one flower is 5-8 days. The blooming period of plants in a population lasts 20-27 days. By its end leaves die off completely, but a flower stalk stays green. Seed ripening takes place during a month, general duration of this period fluctuates slightly and amounts to 15-20 days. Disseminations occurs in May, capsules open and seeds fall. As seeds are large, they turn out to be near the maternal plant. Not all capsules open by the moment of drying up of the flower stalk, their ripening occurs on the substrate after stalk lodging. Additional roots die off, and the plant as an onion enters into dormancy. In autumn, in late October-November in case of precipitation and at temperature higher than +10°C, roots begin to grow, but in contrast of development of typical Mediterranean species the vegetation is absent in *A. trautvetterianum* in autumn and winter and begins only in spring with the onset of stable positive temperatures.

Thus, *A. trautvetterianum* passes the full cycle of seasonal development in the Kulyab Botanical Garden.

Duration of vegetation and phenological phases in introduction depends on weather conditions of the year of observations. So late coming of spring as in 2012-2013 led to a shift of development phases almost for 20 days. The analysis of seasonal development pattern of *A. trautvetterrianum* showed that the growing period in the species under study began with the onset of temperature of 5–10°C. By the nature of vegetation *A. trautvetterrianum* belongs to shortvegetating plants, ephemeroids. Its rhythm of development fully corresponds to the eastern variant of the Mediterrenean climate.

The work was carried out within the framework of the projects of State Assignments ГР № 01011ТД065 (Tajikistan) and № AAAA-A17-117012610053-9 (CSBG SB RAS, Russia).

References

1. A. J. Miller-Rushing, J. Weltzin, New Phyt. **184**, 743-745 (2009)
2. E. Buisson, S. T. Alvarado, S. Le Stradic, L. P. Morellato, Restor. Ecol. **25**, 164-171 (2017)
3. W. T. Stearn, The Kew bot. Mag. **9**, 180-182 (1992)
4. R. M. Fritsch, *A preliminary review of Allium subg. Melanocrommyum in Central Asia* (IPK Gatersleben, Germany, 2016)
5. Z. N. Philimonova, Introdution and Acclimatization. **4**, 91-99 (1966)
6. V. A. Cheryomushkina, Biology of Allium species in Evrasia (Nauka, Novosibirsk 2004)
7. O. V. Daeva, Ecology and Introduction Plants. **9**, 110-114 (1963)
8. L. A. Tukhvattulina, L. M. Abramova *Introduction of wild onions in Bashkortostan: biology, reproduction, agricultural techniques, use* (Ufa, 2012)
9. P. E. Bulakh, *Allium of the natural flora of Central Asia and their culture in Ukraine* (Kiev, 1994)
10. K. Pistrick, *The genus Allium – Taxonomic Problems and Genetic Resources* (Gatersleben, Germany, 1992)
11. T. Boboev, M. Boboev, Sh. Kullaev. *Medicinal and edible plants of Southern Tajikistan, as well as ways of their cultivation and protection* (Kulyab, 2013)
12. I. N. Beydeman, *Methods of studying the phenology of plants and plant communities* (Moskow, 1974)
13. I.V. Borisova, Field geobotany. **4**, 5-94 (1972)
14. I. G. Serebryakov, Bull. Mosc. Soc. of Natur. **69**, 62-75 (1964).
15. N. N. Izmaylova, *Deciduous xerophilic forests, light forests and shrubs* (St-Peterburg, 1995)
16. *Agroclimatic resources of the Tajik SSR* (Leningrad, 1976)

Mycobiota features of aphyllophorales mushrooms in arid conditions of the south-east Euroopean Russia

Nadezhda Kuragina *

Volgograd State University, Departament of biology, 400062, Universitetsky Ave., 100, Volgograd, Russia

Abstract. An analysis of the biota of the aphyllophorales mushrooms carried out in the territory of the Volga-Akhtuba floodplain natural park within the Volgograd region. Based on the results of the conducted researches, 170 species of the study group are noted in the park. The great majority of identified species are saprotrophs on dead fallen wood. The largest number of species was found on *Quercus robur*, *Populus alba*, *P. nigra* и *Fraxinus lanceolata*. The largest number mushrooms ecological groups for moisture are mesophiles and xerophiles, which is typical for the arid zone.

1 Introduction

The natural park «Volga-Akhtuba floodplain» is located between the Volga River and its sleeve Akhtuba, representing the vegetation islet in the middle of plain zone. There are azonal and extrazonal primitive plant aggregation and a peculiar flora of tree and shrubbery plantings unique to the arid South-East of European Russia [1, 2].

Until now, systematic study of the mycobiota of the northern part of the Volga-Akhtuba floodplain (hereinafter VAF) has not been carried out, in the literature there exists fragmentary information about the species composition of the aphyllophorales mushrooms and their ecology features in the region [2-7].

At the same time, it should be emphasized that it is the northern part of the VAF within the Volgograd region that is the most representative part of the VAF with the maximum diversity of landscapes, forest communities and the corresponding flora and mycobiota.

The goal of research is to study the species composition and the ecology of the aphyllophorales mushrooms in the territory of the VAF within the Volgograd region.

2 Materials and methods

The mycobiota VAF research carried out according to classical methods [8-12]. The total number of samples collected was more than 1000.

Species determined using the works of domestic and foreign mycologists [13-22].

* Corresponding author: pipenko87@mail.ru

Mushroom identification carried out on the basis of light microscopy («BIOLAM RFN-11») using a standard set of reagents and electron microscope (scanning electron microscope Versa 3D) (figure 1). Samples tabulated based on the Microsoft Access 2007 application.

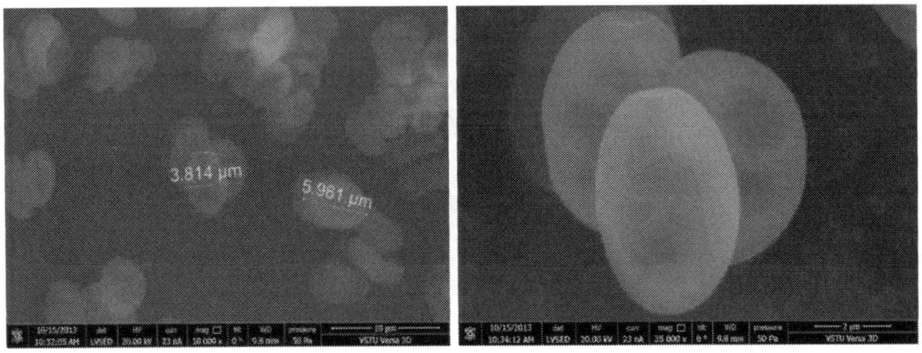

Fig. 1. Electron microscope image of mushroom spores *Cerioporus squamosus* (Huds.) Quél.

3 Results and its discussion

At present, in the territory of the VAF within the Volgograd region 170 aphyllophorales mushroom species were identified. According to the trophic affiliation, more than 90% of all the identified aphyllophorales mushroom belong to saprotrophs, which develop mainly on dead wood (deadwood and fallen trunks, stumps, fallen branches, etc.). Dead fallen wood is one of the most populated mushroom substrates, and some species found also on weakened trees (for example, species such as *Bjerkandera adusta* (Willd.) P. Karst., *Cerioporus squamosus* (Huds.) Quél., *Schizophyllum commune* Fr. etc.), dead trees (*Porostereum spadiceum* (Pers.) Hjortstam et Ryvarden, *Vuilleminia comedens* (Nees) Maire etc.) and on burned wood (*Coniophora arida* (Fr.) P. Karst., *Ganoderma applanatum* (Pers.) Pat., *Phlebia rufa* (Pers.) M. P. Christ. etc.).

For example, species such as *Fomitiporia robusta* (P. Karst.) Fiasson et Niemelä, *Phaeolus schweinitzii* (Fr.) Pat., *Phellinus pomaceus* (Pers.) Maire, *Ph. tremulae* (Bondartsev) Bondartsev et P. N. Borisov, *Pseudoinonotus dryadeus* (Pers.) T. Wagner et M. Fisch. parasitize on living trees and die after their withering away. But mushroom species found also that remain viable after the tree death (*Chondrostereum purpureum* (Pers.) Pouzar, *Fistulina hepatica* (Schaeff.) With., *Ganoderma australe* (Fr.) Pat., *G. lucidum* (Curtis) P. Karst. etc.).

Analyzing the substrate confinement revealed that the prevail number of mushroom is associated with the wood of only broad-leaved trees, the most rich in the species composition of the aphyllophorales mushroom *Quercus robur* L., *Populus* ssp. и *Fraxinus lanceolata* Borkh. (figure. 2). The most species specificity have *Quercus robur*, *Pinus sylvestris* и *Fraxinus lanceolate*. 30 stenotrophic mushroom were recorded on *Quercus robur* (for example, *Abortiporus biennis* (Bull.) Singer, *Daedalea quercina* (L.) Pers., etc.), on *Pinus sylvestris* – 14 stenotrophic species (*Asterodon ferruginosus* Pat., *Auriscalpium vulgare* Gray, etc.), on *Fraxinus lanceolate* – 11 stenotrophic plant (*Ceraceomyces serpens* (Tode) Ginns, *Byssomerulius corium* (Pers.) Parmasto, etc.).

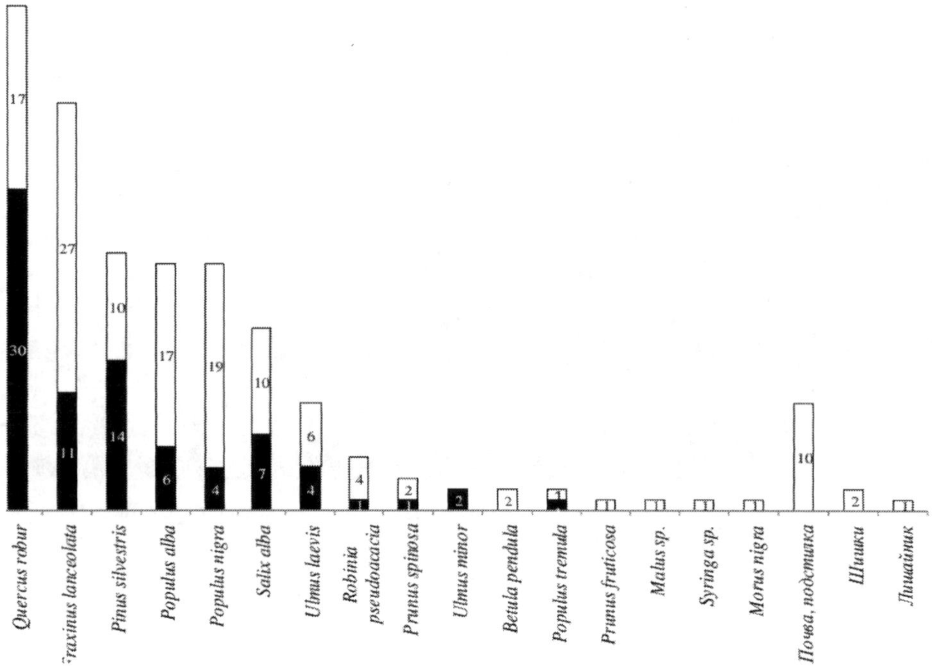

Fig. 2. Confinedness of aphyllophorales mushroom to a different substrate. Note. Above the columns are indicated the total number of mushroom species on this substrate. Black color marked stenotrophic species.

Mycobiota ecological analysis of the northern part of the VAF by moisture showed that mesophiles with a rather high number of xerophiles prevail, which is due to the climatic conditions of the dry-steppe plain zone [23-25].

4 Conclusions

The carried out researches showed that the biota of the aphyllophorales mushroom of the VAF is an original complex, different from the mycobiota of the Volgograd region. The natural park inhabits some mushroom species that determine the analyzed mycobiota originality and can be considered as including habitats of rare mushroom species (for example, *Fistulina hepatica* (Schaeff.) With., *Ganoderma lucidum* (Curtis) P. Karst.) and their complexes, which determines the need to give some of them environmental status [26].

I express my gratitude to Vlasenko V.A. (CSBG SB RAS), Sagalaev V.A. (VolSU) for the consultation, to Volobuev S.V. (BIN RAS) for the assistance in the identification of certain mushroom species, to Golovanova M.A. (VolSU) for the assistance in the collection of mushrooms and many others.

References

1. V.A. Sagalaev, Nat. park «VAF»: nat. resour. potent. 32-68 (2004)
2. V.V. Malychenko (sc.ed. and comp.), *Volga-Akhtuba floodplain is a natural gift to mankind. Illustrated popular science essay on nature protection* (Izdatel, Volgograd, 2006)

3. Yu.A. Rebriev, V.A. Rusanov, T.S. Bulgakov, T.Yu. Svetasheva, I.V. Zmitrovich, E.S. Popov, *Mycobiota of the arid territories of the south-west of Russia* (SFU Publishing, Rostov-on-Don, 2012)
4. N.S. Kuragina, Fundam. res. **(11)**, 211-213 (2013)
5. N.S. Kuragina, Vestnik NSU **12**, 32-36 (2014)
6. N.S. Kuragina, Vestnik VSU **(24)**, 111-116 (2014)
7. N.S. Kuragina, *Mater. IV (XII) Int. Bot. Conf. Young Sci.*. 211 (St. Petersburg, 2018)
8. A.S. Bondartsev, R.A. Zinger *Guide to collecting higher basidiomycetes for scientific research* (Academy of Sciences USSR Publishing, M.-Leningrad, 1950)
9. A.I. Tolmachev, *Introduction to plant geography* (LSU Publishing, Leningrad, 1974)
10. B.A. Yurtsev, Bot. J. **60**, 69-83 (1975)
11. V.M. Schmidt, *Statistical methods in comparative floristics* (LSU Publishing, Leningrad, 1980)
12. V.M. Schmidt, *Mathematical Methods in Botany* (LSU Publishing, Leningrad, 1984)
13. M.A. Bondartseva, E.H. Parmasto, *Families of the Hymenochaetales, Lachnocladiaceae, Coniophoraceae, Schizophyllaceae. Mushroom determinant of the USSR. Order Aphyllophorales* (Science, Sankt-Peterburg,1986)
14. V.A. Mukhin, *A field determinant of pungent mushroom* (UFU Publishing, Yekaterinburg, 1997)
15. M.A. Bondartseva, *The mushrooms determinant of Russia. Order Aphyllophorales* (Science, Sankt-Peterburg, 1998)
16. E.H. Parmasto, *The Determinant of Coral Mushrooms of the USSR* (Science, M.-Leningrad, 1965)
17. J. Eriksson, L. Ryvarden, *The Corticiaceae of North Europe. Hyphodermella-Mycoacia* (Fungiflora, Oslo, 1976)
18. L. Ryvarden, R.L. Gilbertson, *European polypores. Part 1. Abortiporus-Lindtneria* (Fungiflora, Oslo, 1993)
19. L. Ryvarden, R.L. Gilbertson, *European polypores. Part 2. Meripilus*-Tyromyces (Fungiflora, Oslo, 1994)
20. T. Niemelä, *Polyporaceae of Finland and the adjoining territory of Russia* (Publ. University of Helsinki, Helsinki, 2001)
21. T. Niemelä, *Torikseened Soomes ja Eestis (Estonian-language edition of the Finnish original)* (Publ. University of Helsinki, Helsinki, 2005)
22. V.M. Kotkova, T. Niemelä, I.A. Viner, D.S. Schigel, A.V. Kurakov, *Polypore fungi: materials of the international course on ecology and taxonomy of lignicolous basidiomycetes in the Central Forest reserve* (Helsinki University Printing House, Helsinki, 2015)
23. M.A. Safonov, Adv. curr. nat. Sci. **8**, 62-65 (2015)
24. M.A. Safonov, Vestnik OSPU. Electron Sci. J. **3**, 23-31 (2017)
25. A.-L. Ylisimiö, M. Mönkkönen, V. Halikainen, T. Ranta-Maunus, J. Kouki, For. Ecol. Manage. **373**, 138-148 (2016)
26. V.A. Vlasenko, A.V. Vlasenko, Sci. Bull. BSU **41**, 70-78 (2017)

Study of chromosome composition of the southern Ural genotypes of *Prunus pumila* L. by various methods

Mikhail S. Lyozin[1,*], *Sergey* V. Asbaganov[1], *Olga* V. Mochalova[2], *Dmitry* A. Gusev[2], and *Vladimir* S. Simagin[1]

[1]Central Siberian botanical garden of SB RAS, 630090, Novosibirsk, Russia
[2]Lisavenko Research Institute of Horticulture for Siberia, Barnaul, Russia

Abstract. *Prunus pumila* L. is a winter-hardy Siberian species widely used in kitchen-gardens, as well as in nurseries as a seed stock for plums and apricots and development of hybrids for stocks of these cultures. Rare cases of emergence of polyploids among introduced seedlings are known. Chromosome numbers of 11 morphologically different plants of this species from Chelyabinsk Oblast were studied in comparison with some plants from Barnaul with already known ploidy by cytological and cytometric methods. With the help of both methods it was established that all plants from Chelyabinsk Oblast turned out to be diploids. Ploidy of the control specimens from Barnaul was supported by the cytometric method. A rare emergence of polyploids when introducing P. pumila in the climatic conditions differing from the natural ones was confirmed.

1 Introduction

Prunus pumila L. is a North-American species belonging to the section Spiraeopsis Koehne [1].This species has a high frost resistance, and therefore is often grown in Russia in steppificated areas of the Volga regions, the southern Urals and Siberia [2] in the kitchen-gardens and is used in nurseries as a seed stock for plums and apricorts. In this case an intraspecific diversity of seedlings is noted by various morphological features of bushes, leaves and fruits.

As all diploid representatives of this subgenus, except for *Prunus prostrata* Labill., it is characterized by stable diploid (2n=16) set of chromosomes in somatic cells. Aberration is often linked with cultivation in atypical conditions, e.g. when introducing. Spontaneous triploid (2n=24) genotypes are revealed very seldom. I.S. Rudenko managed to find one autotriploid form differing by vegetative heterosis and poor fertility in Moldavian gene pool [3]. One tetraploid form of *Prunus tomentosa* L. (N 14-17 A) was found in the Altaian gene pool among seedlings from the seeds collected by G.V. Eremin in the natural population in Primorsky Krai. Seedlings from free pollination of this form (N63-25-1 and N 63-25-2) had a triploid set of chromosomes [4]. From randomly taken seeds of *Prunus pumila* L. for sowing it was possible to single out spontaneous triploid form N 9 which was later used for

* Corresponding author: lezin-misha@mail.ru

experiments of in vitro artificial polyploidy [5]. Up to now spontaneous autotetraploids for *P.pumila* have not been found.

It should be noted that unlike the representatives of a pure species, intergeneric hybrids of *P. pumila* with diploid plums turn out to be often polyploids, mainly triploids due to functioning of unreduced macrogametes. In the Altaian gene pool such triploids were found among the old genetic materials obtained in the Russian Far East, Krasnoyarsk and the Altai low mountains in the first half of the last century. Such triploids were: SVG 11-19 (a valuable stock for Ussurian plum), Seyanets Yuty, Smena, and Fialka. Sixty three triploids were found in the generations from free pollination of diploid SVG cultivars such as Pchyolka, Chulym, Novinka, Yuta, Opata, Desertnaya, Dalnevostochnaya and Lyubitelskiy [4].

Consideration of interrelation of significant morphological diversity with possible changes of chromosome composition of introduced seedlings is of chief interest. The material of our study were 11 genotypes of *P. pumila* from the nursery of the Company "Gardens of Russia" in southern Chelyabibsk Oblast which differed significantly among themselves in height, habitus, leaf and fruit features. Barnaul *P. pumila* genotypes with known ploidy and triploid hybrid SVG 11-19 were used in the experiment for the comparison. Our aim was study of chromosome composition of these samples by two different methods and search for possible aberrations.

2 Methods

For chromosome counts vegetative buds at the stage of "green cone", previously cleaned from external scales were used. They were pretreated in the mixture (1:1) of 0.002 M 8 –hydroxyquinoline and saturated solution of paradichlorobenzenel for 5 hours [6]. The buds were fixed in 3:1 ethanol: ice-cold acetic acid. The chromosomes were stained by acetic hematoxylin according to the methods of Central Genetic Laboratory [7]. They were counted on not less than 10 metaphase plates using the light microscope Jenaval (Zeiss) equipped with digital camera TC-500.

A ploidy analysis by measurement of nuclear DNA amount was also performed using flow cytometry. Currently this method is widely and successfully used in studying ploidy of various plants, including representatives of the genus Prunus L. [8,9]. Cell nuclei were isolated using buffer CyStain UV Ploidy (Partec) with fluorochrome DAPI (4,6-diamidino-2-phenylindole). All procedures were carried out according to the standard protocol of manufacturer. The measurement of genome size was conducted using FloMax software on the flow cytometer CyFlowSpace (Partec, Germany). More than 5000 cell nuclei were analyzed for each of samples. This present-day method with good technological support allows to study quickly a great quantity of samples.

3 Results

All eleven *P. pumila* forms (N 2-27-51, 2-90-07, 2-59-07, 2-85-07, 2-24-07, B-13-18, 2-25-07, 2-30-07, 2-80-07, 2-18-07, and 2-50-07) for cytological study on the light microscope from the Ural gene pool turned out to be diploids (2n=16). Polyploid cells in the total mass of divisible ones observed on squashed slides were not found. Ploidy of Barnaul *P. pumila* specimens and a remote hybrid, determined earlier, was also confirmed (Table).

Cytometric research fully confirmed the data obtained in cytological study of all experimental genotypes. Despite noticeable distinctions between their morphological features, any changes in ploidy were not found. Both methods may be successfully used, the choice can be determined by available equipment and reagents, as well as by vocational training of specialists.

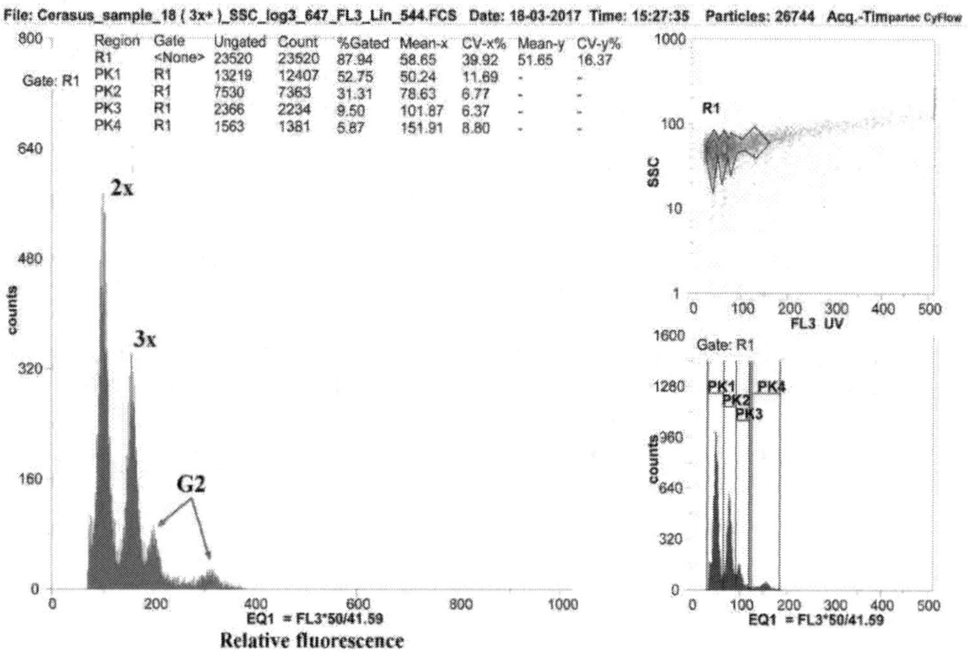

Figure. Histograms of flow cytometry, illustrating ploidy of DNA in the mixture of cell nuclei: diploid "micro P. pumila" 2-85-07 (PK1), triploid "SVG -11-19" (PK2) and corresponding peaks of fuorescence of cell nuclei in phase G2 (PK3 and PK4).

Table. Karyological data of studied specimens and *P. pumila* hybrid

Samples	Specimen Ploidy by data		Chromosome number
	Cytology	Cytofluorimetry	
2-85-07	2x	2x	16
SVG-11-19	3x	3x	24
P. pumila, Barnaul	2x	2x	16
P. pumila, Barnaul	3x	3x	24
2-27-51	2x	2x	16
2-90-02	2x	2x	16
2-59-07	2x	2x	16
2-24-07	2x	2x	16
B-13-18	2x	2x	16
2-25-07	2x	2x	16
2-30-07	2x	2x	16
2-80-07	2x	2x	16
2-18-07	2x	2x	16
2-50-07	2x	2x	16

4 Discussion

Both methods used showed a high effectiveness in determining ploidy. The method of flow cytometry allows to quickly to obtain data on DNA ploidy and the labour consuming traditional one of direct chromosome count should be used for refinement of discovered distinctions. *P. pumila* showed a high cytological stability when being introduced in different conditions compared to those of the natural range, which agrees with the data of Mochalova et al. (202, 2016). It was shown that changes of natural ploidy took place very seldom. Our study confirms that it is unlikely to find such genotypes by the deviations from morphological features typical of the species. To get similar changes, one should apply diverse special actions on the genotype, including previous obtaining polyploids and involvement of them in hybridization or conducting remote crossings. The karyological data obtained on *P. pumila* specimens may be used in future in introduction and breeding, including formation of polyploid genotypes.

References

1. G.V. Eremin, *Pomology, Systematics of stone fruit plants* (Orel, 2008) in Russian
2. A.B. Gobunov, *Introduction of non-traditional fruit, berry and vegetable plants in Western Siberia* (GEO, Novosibirsk, 2013) in Russian
3. I.S. Rudenko, *Remote hybridization and polyploidy in fruit plants* (Chişinău, Shtiintsa, 1978) in Russian
4. O.V, Mochalova, D.A. Gusev, Ach. of sc. and tech. of agr. **30**, 36-39 (2016) in Russian
5. O.V, Mochalova, M.N. Matiunin, *Cytoembryology and selection of remote hybrids and stone plant polyploids in Altai* (SB RAS, Novosibirsk, 2002) in Russian
6. Z.P. Pausheva, *Workshop on plant cytology* (Agropromizdat, Moskow, 1988)
7. *Cytological studies of fruit and berry crops / Guidelines* (CGL, Michurinsk, 1976) in Russian
8. M. Zabka, L. Durisova, P. Elias Jr, T. Baranec, Biologia **73**, 121–128 (2018)
9. W.V. Baird, A.S. Estager, J.K. Wells, J. Amer. Soc. Hort. Sci. **119**, 1312–1316 (1994)

Assessment of genetic fidelity of *Fritillaria dagana* (Liliaceae) regenerated plants using ISSR markers

Dinara S. Muraseva[1,*], *Elena V.* Kobozeva[1,2], and *Tatyana I.* Novikova[1]

[1]Central Siberian Botanical Garden, Siberian Branch RAS, 630090 Novosibirsk, Russia
[2]National Research Tomsk State University, 634050, Tomsk, Lenina pr., 36

Abstract. ISSR analysis of *Fritillaria dagana*, endangered ornamental geophyte from the Sayan Mountains, regenerated through direct gemmogenesis from bulb scale tissue was performed. More informative electrophoresis profile with clear and distinct bands was obtained at amplification with $(CAC)_3GC$ primer at 56 °C annealing. The genetic fidelity of *F.dagana* regenerants to maternal plants was confirmed.

1 Introduction

The genus *Fritillaria* L. (Liliaceae) presents ephemeral bulbous plants and comprises more than 150 species distributed within the Northern Hemisphere with a pronounced center of diversity in southwestern and Himalayan Asia [1]. Twelve of fritillaria species occur in the territory of Russia, therewith five of them are listed in the Red Data Book of the Russian Federation [2] with the status of rare species. Since many natural species of the genus *Fritillaria* are endangered due to irregular gathering the bulbs and flowering shoots, it is necessary to find new effective approaches for their conservation. Previously we developed the protocol of clonal micropropagation of local endemic *Fritillaria sonnikovae* Schaulo and A. Erst from West Sayan for creation of *in vitro Fritillaria* collection [3].

Fritillaria dagana Turcz. ex Trautv. is an endemic species listed in the Red Data Book of the Russian Federation [2] with the status of a rare species (3a) which occurs in certain areas of Siberia, East Sayan and Southern Baikal. An efficient micropropagation protocol of *F. dagana* was developed for the first time [4].

Although clonal propagation should generate plantlets identical to the mother plants, studies have shown that genetic and epigenetic changes called somaclonal variations may occur due to callus formation, use of growth regulators and long period of cultivation. Therefore the risk of somaclonal variation appearance should be assess by using various methods including morphological, biochemical and molecular ones [5, 6]. Molecular methods represent an effective tool for detection of regenerant genetic variability, which is explained by a higher level of DNA changes in comparison with morphological polymorphism. One of the effective methods required in this study is the ISSR analysis (Inter Simple Sequence Repeats). This method is easy to use, low-cost, high reliable,

* Corresponding author: dsmuraseva@csbg.nsc.ru

reproducible, and methodologically less demanding compared to other dominant markers [7, 8, 9]. To date, there is a large number of studies devoted to the detection of somaclonal variation using the ISSR analysis [10, 11, 8].

The present study was undertaken to assess somaclonal variation with the use of ISSR markers in *F. dagana* regenerants obtained through direct gemogenesis from bulb scale tissue.

2 Material and methods

The aseptic scales were cut into the segments of 5 × 5 mm size and used as primary explants. Scale segments (for 4–5 pcs.) were placed by cut-surface down onto induction medium. The cultivation was carried out on solidified B_5 medium [12] supplemented with 5.0-μM 6-benzylaminopurine (BAP) and 2.0-μM α-naphthalene acetic acid (NAA). Microclones of *F. dagana* obtaned through direct regeneration from bulb scale tissue were used for genetic analysis. *In vitro*-raised microbulbs (regenerants) and maternal bulb scale segments were analysed.

DNA was extracted from silica dried bulblets using a Food and Raw materials extraction DNA Kit (BioSilica, Russia) and purified on columns following the manufacturer's protocol. Purity and concentration of the extracted DNA were evaluated using BioSpectrometer kinetic (Eppendorf, Germany) spectrophotometer with μCuvette G1.0 (Eppendorf, Germany) microcuvette.

Fifteen ISSR primers were initially screened: $(CA)_6GT$, $(CA)_6AG$, $(CT)_8TG$, $(AG)_8YC$, $(CAC)_3GC$, $(CTC)_3GC$, $(AC)_8CG$, $(AC)_8YG$, $(GA)_8YC$, $(CT)_8AC$, $(CA)_6AC$, $(CA)_6GG$, $(CA)_6RG$, $(GAA)_6$, $(GACAC)_4$. PCR was performed on C 1000 Thermal Cycler (BioRad Laboratories, USA). The reaction mixture of 15.0 □L contained 2.7 □M $MgCl_2$, 1.25 □M primer, 0.4 □M dNTP, 1× PCR buffer, 1.5 of Taq polymerase (Medigen, Russia) and 5.0 ng of template DNA. The amplification reaction consisted of an initial denaturation step at 94 °C for 1.30 min, followed by 35 cycles of 40 s denaturation at 94 °C, 45 s annealing at 51 °C and 56 °C, 1.30 min extension at 72 °C with a final extension of 72 °C for 5 min. The amplified products were separated by gel electrophoresis on 1.5 % agarose gels in 1×TBE buffer and stained with SYBR-Green (Medigen, Russia). The sizes of the amplification products were estimated with a DNA ladder (Medigen, Russia). DNA fragments were visualized using Gel Doc XR+ and analyzed with Image Lab Software (BioRad Laboratories, USA).

3 Results and discussion

The first changes on the scale surface – the overgrowth of explant tissue – were observed 55–56 days after inoculation on nutrient media. No formation of callus on the *in vitro* culture initiation stage was noted, bulblet regeneration occurred through direct gemmogenesis. The emergence of buds was occurred on the undamaged part of bulb scale protruding above the surface of the medium, but growth of them – close to the wound surface. The frequency of the microshoot regeneration achieved 66 % and 3.6 ±0.4 bulblets were formed, on the average, per one explant. Adventitious bulblets obtained from direct regeneration were separated from the primary explant and used for DNA extraction.

At screening of fifteen ISSR primers, only six were effective (Table). At the stage of the preliminary experiment we tested two annealing temperatures – 51 °C and 56 °C. As a result of DNA-PCR amplification, 5–17 bands ranging in size from 200 bp to 2500 bp depending on the annealing temperature were obtained.

Table The general characteristics of the ISSR primers used in the somaclonal variation tests of *Fritillaria dagana*

Primer sequence (5'-3')	Number of bands, annealing temperature (°C)		Band size, bp
	51 °C	56 °C	
$(CT)_8TG$	8	9	500–1500
$(CAC)_3GC$	14	15	200–2500
$(CTC)_3GC$	14	17	200–2300
$(CA)_6GT$	12	15	200–2500
$(CA)_6AG$	8	14	250–2000
$(AG)_8YC$*	8	14	450–2000

* Y = C or T

Low differentiation of the amplification products was established at 51 °C of annealing for $(CT)_8TG$, $(CA)_6GT$, $(CA)_6AG$ and $(AG)_8YC$ primers, what complicated the analyze of the amplification profile (Fig. 1 a). Unclear bands were observed with $(CA)_6AG$ and $(AG)_8YC$ at 56 °C annealing. The maximum number (17) of bands was obtained using $(CTC)_3GC$. More informative amplification profiles with clear and distinct bands were at PCR with $(CAC)_3GC$ at 56 °C of annealing (Fig. 1 b).

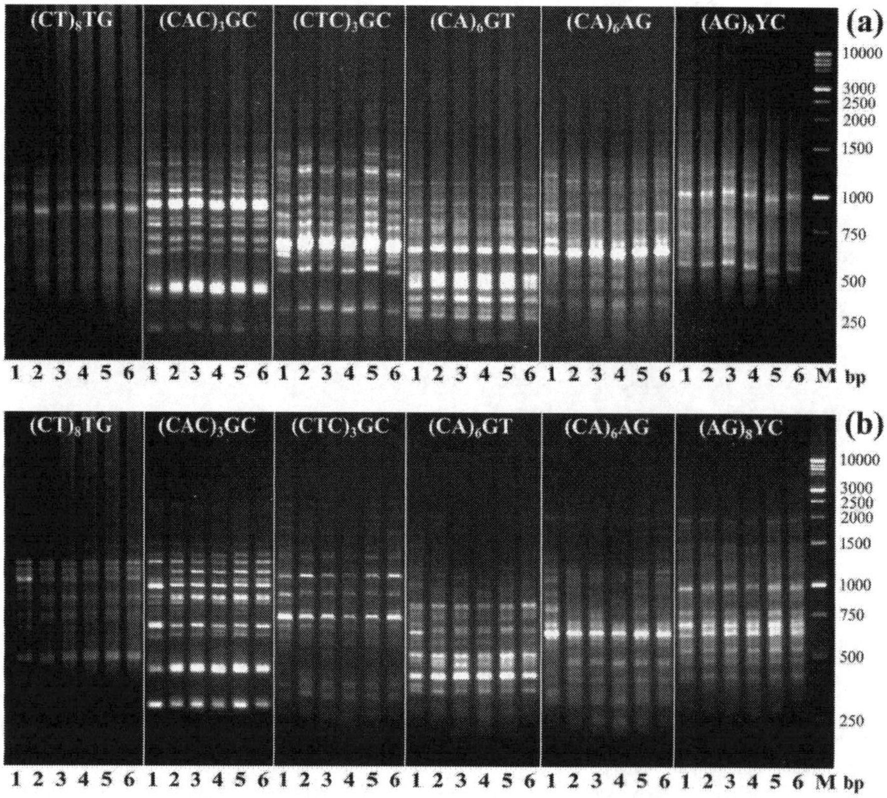

Fig. 1. PCR amplification profiles obtained with ISSR markers of the first generation of *Fritillaria dagana* regenerated plants (2–6 treks) in comparison with the maternal plant (1 trek); primers: $(CT)_8TG$, $(CAC)_3GC$, $(CTC)_3GC$, $(CA)_6GT$, $(CA)_6AG$, $(AG)_8YC$, annealing temperature: **a** – 51 °C, **b** – 56 °C, M – a molecular marker; B_5 nutrient medium supplemented with BAP 5.0 μM and NAA 2.0 μM

Generally, the banding profiles from micropropagated plants (2–6 treks) were monomorphic and similar to those of the mother plants (1 trek). In some cases, the differences between the patterns were observed because of different degree of the amplification products during PCR, wich was clarified and checked by repeatability of PCR. As a result, the genetic fidelity of the regenerants to the maternal plant was revealed.

Genetic marking technologies are widely applied methods for monitoring of genetic fidelity of regenerated plants. The diagnostic capabilities of the ISSR markers used by us were successfully illustrated by numerous studies of somaclonal variation in different plant species [13, 14].

In our study, we relied on the work based on assessment of genetic diversity of *F. thunbergii* Miq. [15] and *F. imperialis* L. [16] nature populations by ISSR analysis. We used two primers – $(AG)_8YC$ and $(GAA)_6$ which were offered in the publications. However, in the work with *F. dagana* regenerants the primers didn't show the expected polymorphism and were not considered in further analysis of somaclonal variation. Perhaps the reason is that these species belong to different subgenera – *Fritillaria, Petilium* and *Liliorhiza,* respectively. Previously we used the ISSR markers for identifying somaclonal variation of *F. sonnikovae* Schaulo et A. Erst after *in vitro* slow-growth storage [3]. It was established that the genetic variation process of regenerated plant formed through direct organogenesis did not occur at the first passage following *in vitro* storage for twelve months. We applied the same primer set and achieved the informative PCR amplification profile. The similarity of the effectivness of the ISSR primers can be explained by the position of this plants in one intraspecies group – subgenera *Liliorhiza.*

Thus, the results allow concluding that there is no somaclonal variation of *F. dagana* regenerated plants obtained within the direct gemmogenesis from bulb scale explant tissue. However, these results should be considered preliminary because the ISSR analysis for *F. dagana* was applied for the first time. In future it is planned to optimize the DNA extraction protocol, PCR program, and to carry out a statistical data processing for a large number of primers wich will ensure reliability of the results.

The reported study was funded by RFBR according to research project № 18-34-00164\18.
In our study, material from the collection of the Central Siberian Botanical Garden SB RAS – USU_440534 "Collection of living plants indoors and outdoors" was used.

References

1. N. Rønsted, S. Law, H. Thornton, M.F. Fay, M.W. Chase Mol. Phylogen. Evol. **35,** 509–527 (2005)
2. L.V. Bardunov, V.S. Novikov, *Red Book of Russian Federation (plants and fungi)* (KMK Scientifc Press Ltd., Moscow, 2008)
3. D.S. Muraseva, N.S. Zvyagina, T.I. Novikova, O.V. Dorogina, Vavilov J. Gen. Breed. **21**, 554–560 (2017)
4. A.A. Erst, A.S. Erst, D.N. Shaulo, D.S. Kulkhanova, Flora Asian Rus. **1**, 64–70 (2014)
5. V. Rani, A. Parida, S. Raina Plant Cell. Rep. **14,** 459–462 (1995)
6. O.M. Bublyk, I.O. Andreev, K.V. Spiridonova, V.A. Kunakh Biol. Plant. **56**, 459–462 (2012)
7. M.W. Bairu, C.W. Fennell, J. Van Staden, Sci. Hortic. **108**, 147–173 (2006)
8. X. Liu, G. Yang, In Vitro Cell. Dev. Biol. – Plant. **48,** 172–179 (2012)

9. W. Al Khateeb, E. Bahar, J. Lahham, D. Schroeder, E. Hussein, Physiol. Mol. Biol. Plants. **19**, 157–164 (2013)
10. S.S. Das Bhowmik, A. Basu, L. Sahoo, J. Crop Sci. Biotech. **19**, 157–165 (2016)
11. S. Saha, S. Adhikari, T. Dey, P. Ghosh, Meta Gene **7**, 7–15 (2016)
12. O.L. Gamborg, D.E. Eveleigh, Can. J. Biochem. **46**, 417–421 (1968)
13. Z.-F. Yin, B. Zhao, W.-L. Bi, L. Chen, Q.C. Wang, In Vitro Cell. Dev. Biol.—Plant **49**, 333–342 (2013)
14. M.A. Ramírez-Mosqueda, L.G. Iglesias-Andreu, Plant Cell Tiss. Organ. Cult. **123**, 657–664 (2015)
15. S. Li, K. Hu, J.Guo, X. Yang, Y. Zhu, Z. Cheng, Biochem. Syst. Ecol. **39,** 725–731 (2011)
16. S. Badfar-Chaleshtori, B. Shiran, M. Kohgard, H. Mommeni, A. Hafizi, M. Khodambashi, N. Mirakhorli, K. Sorkheh, Biochem. Syst. Ecol. **42**, 35–48 (2012)

The results of the introduction of *Juglans mandshurica* Maxim. into the oak-woods of the Khopyor region

Anastasia Ovcharenko [1,*], and *Alevtina* Ovcharenko [1]

[1] Balashov Institute of Saratov State University, 412300, Balashov, str. Karl Marx, 29, Russia

Abstract. An analysis of the results of the introduction of *Juglans mandshurica* Maxim. into the flood-plain oak-woods of the Khopyor is conducted. The total bioecological assessment and prospects of the introduction into the natural flood-plain oak-woods of the Middle Khopyor region for the enrichment of their species composition with the aim of enhancing the sustainability of these ecosystems and optimization of their ecological functions are given. The compliance of the environmental conditions of the growth of introduced stands in the middle reaches of the Khopyor River, original within the natural distribution area is found. Alongside with the identified typical directions of reduction in the indicators of the stability and productivity of this breed in introduced centres in comparison with the natural areas there detected trends in increasing the resistance of the species in the succeeding generations, as a result of clone variability, a selection of species with a high vital potential and adaptive capacity.

1 Introduction

In the modern issues, an important place is given to the optimization of natural oak plantations. Within the framework of solving this problem, there are tasks to increase the productivity of plantations, enrichment of aboriginal flora with economically valuable species of trees and shrubs. One of the directions of forestry of the last century was the introduction of valuable species of wood introducents into forest crops. Zoning and acclimatization of species from other regions expands the diversity of local flora, increases the biological potential. The endurance of woody plants of exotic species is the most objectively revealed in the course of a long introduction tests.

The purpose of these studies was to study the current state of the *Juglans mandshurica* Maxim. plantation, introduced in the mid-twentieth century in the natural forest the Middle Khopyor region.

2 Materials and Methods

* Corresponding author: ow4arenko.nastia@yandex.ru

A grove of *Juglans mandshurica* Maxim. was planted by a forestry of the Balashov in the Pads precinct forest area near the village of Kotovras in the West of Saratov region in a natural grove of the Middle Khopyor region. This grove was announced by the decision of the Saratov regional Executive Committee № 328 of December 27, 1991 as a natural monument "Manchurian walnut Grove". This is one of the two in the field of seed areas of the far Eastern nature – the Manchurian nut – at the age of 65 years. The nature monument has regional, scientific-genetic and ecological significance. In the autumn of October 9, 2016 we conducted a survey of the state of this plant.

Features of the climate of the area are dryness, high continental and high variability from year to year, especially in terms of rainfall. It should be noted the unpredictability of the conditions of the vegetation period, which can be both wet and cool and dry, and the timing of the onset, duration and strength of the drought are also changed in a number of years [1].

Methods of field research included conducting standard forest mensuration and geobotanical descriptions at the site, the study of the peculiarities of forest regeneration processes and the characteristics of the environmental conditions of the environment under the canopy of the forest. The vertical structure of phytocenosis was analyzed on all levels: stand, undergrowth and ground cover. For trees, in addition to the usual morphometric parameters (height, trunk diameter at breast level), the life state was also evaluated by visual criteria of the scale V. A. Alekseev [2].

Winter hardiness was determined by 8-point scale S. Y. Sokolov. The ability to resist the harmful effects of drought was determined visually by the degree of damage to plants on a 7-point scale of I. F. Gritsenko. To assess the adaptive capacity of the species, the scale of A. A. Kalinichenko was used in modification Of N. B. Baranovskaya. Assessment of viability and prospects of introducents is carried out by the method of p. I. Lapin [3]. Mathematical processing of results was performed on a PC using the software packages EXCEL and STATISTIKA.

3 Results

Grove is a single plot planting in 1956 with an area of 0.2 hectare. At the present time 95 trees *Juglans mandshurica* Maxim. have height up to 20 meters and a diameter of 20 to 24 cm.

In the plantings trees retain their vital form; almost all have reached the generative maturity. Nut at the age of 60 years has an average height of 18 m, exceeding the indicators in other introduction points in Ukraine, Belarus, Saratov, but not reaching the maximum for the trees of the age limit in nature. The average diameter of the trunk is slightly reduced – 22.75±0.21, which indicates that local conditions are not optimal. The plants of the genus *Juglans* are grown in many introduction centers, but the indicators of their stability and productivity are reduced everywhere in comparison with natural habitats [3].

Nut is a light-loving breed. Stands have rather low closeness – 0.5-0.6. The elm (*Ulmus glabra* Huds) grows evenly over the entire study area.), pine trees are isolated (*Pinus silvestris* L.). When planting, probably, an analogue of the phytocenosis of natural growth of this species – mixed coniferous-broadleaved forests of the southern part of the Far East was modeled [4]. Also, a significant number of individuals of Holly maple (*Acer platanoidec* L.) and ash-tree maple (A. *negundo* L.) were found during the examination. The latter is a weed woody plant and indicates an intensive anthropogenic impact.

The second tier, sparse, up to 6-7 m high from the maple of the Tatar (*Acer tataricum* L.), hawthorn (*Crataegus ambigua* C. A. Mey), buckthorn (*Frangula alnus* Mill.), blackthorn (*Prunus spinosa* L.), spindle tree (*Euonymus verrucosa* Scop.). Two elderberry bushes (*Sambucus racemosa* L.) were met singly. Together with the nut was planted in a significant amount of hazel (*Corylus avellanal* L.), which is not found in the natural forests of the Saratov prihopera. The projective cover of the ground cover is uneven, from 20-30% under a thick woody canopy to 90% in open areas, it is formed from a large number of weeds, forest and weed species that signal violations of the forest habitat of valuable woody plants.

List of stands with their division into classes of life state (Fig. 1) showed his extremely uneven distribution of on stages thickness. The index of the forest stand life state is very low – 17.45. The extremely unfavorable state of life of wood plants, dry-runity was revealed, in some places because of what there is a decrease in the closeness of crowns to 0.3. Probably the poor condition of introduced organisms is the consequences of the several very hot years, and especially the abnormally hot summer of 2010, Drought tolerance is one of the main criteria of successful growth of exotic species in the steppe conditions, in which nut in the planting gets III points during the dry season the leaves wilt, the lower leaves of the shoots take autumn color and fall off. Perhaps its state in the area of introduction is limited by the conditions of water supply and depends on the degree of frequency of dry periods, which is reflected in the graph in the form of dips in some steps of thickness.

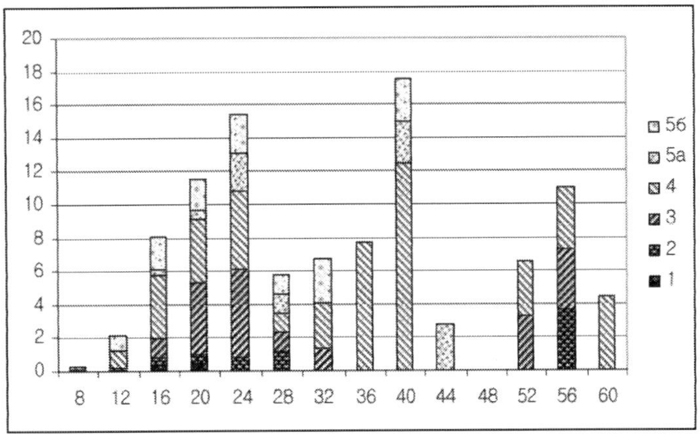

Fig. 1. The distribution of *Juglans mandshurica* Maxim. in the planting of the steps of thickness (abscissa axis, cm) and stock (ordinate axis,%) with the division into classes of living conditions (1-healthy individuals, 2-weakened, 3 – sick, 4 – dying, 5 – dry (5A – fresh dead – trees that died less than a year ago, 5B – old dead – trees that died in the past).

4 Discussion

The success of the introduction largely depends on the compliance of the environmental conditions of the new habitat with the biological characteristics of the species [3]. Manchurian nut grows wild mainly in the valleys of rivers and streams in the Amur region and Primorye, demanding the fertility and humidity of the soil, it grows best on a fairly deep fresh and moderately moist carbonate loams with close occurrence of flowing groundwater. Do not tolerate the wetlands soils [4].

Introduction planting of nut is located in a high part of the Central floodplain of Khoper, flooding in this area is often absent or short-term with a large spill. Type of forest conditions D2P - oak forest floodplain fresh. Thus, the area of the floodplain forest in the middle reaches of the river hopper is quite suitable for its environmental parameters for the experiment. Hydrological conditions of growth coincided with the original within the natural area of distribution. The site was valuable for the scientific study of the possibility of introduction of exot in forest crops of the region.

It should be noted that in the conditions of the floodplain of hopra the Manchurian walnut is rather winter-hardy. This is the most frost-resistant species in the genus, on its trunks during the examination did not meet frost – I class of winter hardiness – overwinter without damage. It blooms early and as a result late spring frosts may do damage.

A special indicator of the state of plants-their resistance to pests and diseases – is closely related to their winter hardiness and drought resistance. Found often instances of the class III the state of life – patients (20,3%) or weakened (6,4%), with various Fitopatologiya, with peeling bark and some dry branches, struck by the various types of tinder, stem phytophages. It should be noted the General disadvantage of the plant-accompanying rocks and undergrowth crops are also weakened, warm and humid weather of the growing season in 2011 contributed to the mass development of root rot pathogens here – autumn shedding (*Armillariella mellea* Quel.), as well as black leaf spot (*Rhytisma acerina* (P.) Fr.), which is found even on the leaves of the nut, although it is believed that it is quite resistant to rot and little damage by insects (except old trees) [4].

The predominance of class IV trees – dying (48.9%) and V – dry (22.8%) indicates the instability of planting at this time. The number of trees that died less than a year ago (9.5%) is approximately the same as the old dead wood on the site. A significant percentage of subsenile and senile individuals is due, in addition to natural and climatic factors, to earlier aging of trees in unusual conditions [3].

The loss of certain groups of the composition of stands in different time periods is illustrated in figure 1 in the form of failures in the degrees of thickness.

Against the background the oppressed living condition of walnut intensified coppice shoot formation from dormant buds at the surface of the stand. Almost all trees bear fruit, forming a benign seeds which with the passage of the stratification to germinate well. But it should be especially noted the absence of self-seeding and generative undergrowth in the planting, on the basis of which this species can be attributed to the III level of adaptation of generative development, an invasive inactive group.

In the model the imposition of most instances of grade 1 vitality – healthy – presented low levels of thickness. These are young vegetative plants formed by dying individuals, which often give a large number of shoots. The diagram clearly identifies the generation of the second and third generation with the maximum of 40 and 24 steps of thickness. The distribution into life classes in them reflects the increase in species resistance in subsequent generations as a result of clonal variability, selection of species with increased life potentials and high adaptive capacity. According to the literature data in genetically isolated populations of introducents, when the generations of 3–4-generation reproducers change in the rhythm of seasonal development, they approach native plants, as a result of phenotypic variability, the adaptation of the phenotype of woody plants to these conditions is completed. Already for 8–10 shifts of generations in introduced plants there is a restructuring and genotype, allowing them to adapt to new climatic conditions [3].

The General unfavorable state of forest plantations is associated with both sharp climatic fluctuations and anthropogenic impact. Directly on the territory of the plant is a working sawmills, other farm buildings, agricultural machinery forestry. It is necessary to ensure the protection and restoration of the valuable site by seed planting.

References

1. L.M. Kavelenova, S.A. Rozno, Bulletin of the Samara State University. Extraordinary issue, 156-165 (2002)
2. V.A. Alekseev, Forest Science J. **4**, 51-57 (1989)

3. A.K. Polyakov, *Introduction of woody plants in the conditions of anthropogenic environment* (Donetsk Botanical Garden of NAS of Ukraine, Donetsk, 2009)
4. N.E. Bulygin, V.T. Yarmishko, *Dendrology* (EDP Sciences, St. Petersburg, 2000)

Karyotypic variability in some species of the genus Chondrilla (Asteraceae)

Alena Parkhomenko[1,*], *Aleksandr* Kashin[1], and *Lyudmila* Grebenyuk[1]

[1]Saratov State University, 83 Astrakhanskaya Street, Saratov, 410012, Russia

Abstract. Karyotypic variability of plants was evaluated in 17 populations of six species of the genus *Chondrilla* (*C. ambigua* Fisch., *C. brevirostris* Fisch, et Mey., *C. laticoronata* Leonova, *C. canescens* Kar., Kir., *C. juncea*, *C. pauciflora* Ledeb.) in the Astrakhan, Voronezh and Saratov regions, the Republic of Kalmykia, and Western Kazakhstan. It is maintained that *C. ambigua* is a strict diploid (2n = 2x = 10) species, while its close relative - *C. pauciflora* - is a strict triploid (2n = 3x = 15) taxon. The research demonstrates that the plants of the apomictic taxa *C. brevirostris*, *C. canescens*, *C. juncea* and *C. laticoronata* are characterized by the genomic instability in the form of chromosome instability which results in these species' mixed-ploidy populations.

1 Introduction

The research on genetic polymorphism and morphological variability of a number of *Chondrilla* species in European Russia [1] requires a thorough examination of the species' kariotypic variability.

2 Materials and Methods

2.1 Object of study

In order to evaluate the karyotypic variability of plants, 17 populations of six *Chondrilla* species were subject to the research. Below are the names of the species, the numbers of the populations (in brackets), and the populations' locations:

C. ambigua (1030) – Astrakhan region, Krasnoyarsky district, near the village of Dosang; (1031) – Republic of Kalmykia, Yashkulsky district, near the village of Khulhuta.

C. brevirostris (1029) – Astrakhan region, Akhtubinsky district, near the village Bolhuny; (1046) – Astrakhan region, Kharabalinsky district, near the village of Volnoe; (1055) – Astrakhan region, Kharabalinsky district, near the station Verblyuzhya; (1058) – Astrakhan region, Krasnoyarsky district, near the village of Dosang; (1059) – Republic of Kalmykia,

* Corresponding author: parkhomenko_as@mail.ru

Yashkulsky district, near the village of Khulhuta; (1077) – Kazakhstan, West Kazakhstan region, Bokeyorda district, near the village of Giekkum.

C. canescens (1040) – Saratov Region, Khvalynsky district, near the town of Khvalynsk, Belenkaya Mountain; (1054) – Astrakhan region, Akhtubinsky district, near the station Kapustin Yar; (1074) – Voronezh region, Bogucharovsky district, near the village Tereshkovo.

C. juncea (1027) – Saratov region, Krasnokutsky district, near the village of Dyakavka.

C. laticoronata (1056) – Astrakhan region, Akhtubinsky district, near the village of Verblyuzhye; (1070) – Astrakhan region, Kharabalinsky district, near the village of Sasykoli; (1071) – Astrakhan region, Kharabalinsky district, near the village of Tambovka; (1076) – Kazakhstan, West Kazakhstan region, Bokeyorda district, near the village of Giekkum.

C. pauciflora (1078) – Kazakhstan, West Kazakhstan region, Bokeyorda district, near the village of Urda.

The achenes that served as the material were collected from the natural populations in the years 2015-2017. The research started with examining the populations of those taxa that had been proved independent species by the earlier research on their morphological variability; and of those two taxa (*C. canescens* and *C. juncea*) the independance of which has not yet been established [1].

2.2 Methods

In order to evaluate the karyotypic variability of plants in the populations, the squash technique was employed to measure the chromosome number in the cells of seedlings' root apical meristems [2]. The chromosome number was calculated in root apical meristems of germinated seeds that had been formed in two flowering regimes (free flowering and dust-free flowering).

The seeds were germinated until the length of roots reached 10-12 mm. The root tips were pre-treated with an aqueous solution of bromonaphtaline for 3.5 hours at 40C. Roots were then fixed in a solution of 3 part 95% ethanol and 1 part 45% acetic acid. The fixed material was stained in aceto-haematoxylin; and the slides were prepared using the squash method in an 80% cloral hydrate solution.

The slides were analysed using an Axio Scope A1 microscope (Carl Zeiss) with immersion oil under 100×16 magnification. The microimages were taken with a high-definition digital camera AxioCam MRc 5 (D) and an adapter 60N-C 1" 1.0x. Chromosomes were couted in no less than 10 cells of each slide [3].

Whenever a plant was diagnosed with aneu- or mixoploidy, its basic ploidy level was considered the one that was characteristic of the majority of apical meristem cells.

3 Results

The basic chromosome number $x = 5$ was used to determine the ploidy level [4, 5]. It is revealed that only two of the studied taxa are constituted of plants of the same ploidy level. Specifically, all *C. ambigua* plants are diploid ($2n = 2x = 10$), while all plants of its close relative – *C. pauciflora* – are triploid ($2n = 3x = 15$). These results are in line with the data recorded in the literature [6].

C. juncea plants from the Krasnokutsky district in the Saratov region were either triploid or tetraploid. The majority of *C. canescens* plants were also triploid, with about 10% of diploid plants collected in the population in the Khvalynsky district of the Saratov region. Regardless of the location, *C. laticoronata* plants were triploid, with only a small portion of pentaploid plants in one of the populations in the adjacencies of the Tambovka village in the Astrakhan region.

Table 1. Ploidy levels in the root apical meristem cells of *Chondrilla* under different flowering regimes in 2015-2017.

Species name	№ pop.	Year of study	Flowering regime	Analyzed seedlings					
				total, units.	of them %				
					x	2x	3x	4x	5x
C. ambigua	1030	2015	f/f	3	0	100	0	0	0
		2016	f/f	3	0	100	0	0	0
		2017	f/f	36	0	100	0	0	0
	1031	2015	f/f	4	0	100	0	0	0
C. brevirostris	1029	2015	f/f	8	0	0	12.5	87.5	0
		2016	f/f	9	0	0	100	0	0
			d/f	3	0	0	100	0	0
		2017	f/f	47	0	0	100	0	0
	1046	2016	f/f	62	0	0	100	0	0
	1055	2016	f/f	24	0	8.3	87.5	4.2	0
		2017	f/f	19	0	0	100	0	0
	1058	2016	f/f	10	0	0	90.0	10.0	0
			d/f	2	0	0	100	0	0
	1059	2016	f/f	61	1.6	98.4	0	0	0
	1077	2017	f/f	71	0	0	100	0	0
C. canescens	1040	2016	f/f	54	0	9.3	90.7	0	0
	1054	2016	f/f	12	0	0	100	0	0
	1074	2017	f/f	5	0	0	100	0	0
C. juncea	1027	2015	f/f	11	0	0	72.7	27.3	0
			d/f	2	0	0	50.0	50.0	0
		2016	f/f	19	0	0	52.6	47.4	0
C. laticoronata	1056	2016	f/f	39	0	0	100	0	0
	1070	2016	f/f	10	0	0	100	0	0
	1071	2016	f/f	2	0	0	100	0	0
		2017	f/f	35	0	0	94.3	0	5.7
	1076	2017	f/f	53	0	0	100	0	0
C. pauciflora	1078	2017	f/f	19	0	0	100	0	0

Note. x=5. Flowering regime: f/f – free flowering, d/f – dust-free flowering.

Of all the studied populations, *C. brevirostris* had the highest degree of ploidy variation. In two out of six populations of this species (populations no. 1046 and 1077 of the Astrakhan region and Western Kazakhstan respectively) plants had stable ploidy 3x=15. In other populations ploidy ranged from x=5 (one plant of population no. 1059) to 4x=20 (7 plants of population no. 1029 of the year 2015 and rare plants of populations no. 1055 and 1058 of the year 2016).

Furthermore, a detailed analysis revealed a significant proportion of aneu- and mixoploid plants among *C. brevirostris*, *C. canescens*, *C. juncea*, and *C. laticoronata*. Also, for these species one and the same apical meristem may contain cells with different ploidy levels and diverse variants of aneuploidy. It is only in *C. ambigua* and *C. pauciflora* plants where aneuploidy and mixoploidy never occurs.

4 Discussion

The research revealed the genomic instability in the form of chromosome instability in *C. breverostric*, *C. canescens*, *C. juncea*, and *C. laticoronata* plants. Earlier these species were found to have notable variability in the frequencies of gametophytic apomixis in the populations [7]. The data suggest that in these species the variability of apomixis frequency results from the genome instability in apical meristem cells, while the genome instability itself comes from the hybridisation of the studied taxa or their interbreeding with other species of the genus. The strict genome diploidy of *C. ambigua* asserts that the species is obligate apomictic. Although the *C. pauciflora*'s capacity for apomictic seed formation has not been investigated, the strict genome triploidy of its plants provides evidence that it is an apomictic taxon and, most likely, a cytotype deriving from *C. ambigua* which it has very similar morphological [8] and molecular and genetic [9] features with. Apparently, this cytotype is not linked to the parental species by gene flow and is obligate apomictic.

This study was supported by the RFBR (project no. 15-04-04087).

References

1. A. Kashin, T. Kritskaya, A. Popova, A. Parkhomenko, Bulletin of Moscow Society of Naturalists. Biological series **122**, 50-70 (2017).
2. L. Abramova, *Determination of the number of chromosomes and a description of their morphology in meristems and pollen grains of cultivated plants* (Leningrad, 1988).
3. V. Grif, N. Agapova, Botanicheskii Zhurnal **71**, 550-553 (1986).
4. L. Bolkhovskikh, V. Grif, O. Zakharyeva, T. Matveyeva, *Chromosomal numbers of flowering plants* (Nauka, Leningrad, 1969).
5. N. Agapova, K. Arkharova, L. Vakhtina, Ye. Zemskova, L. Tarvis, *Number of chromosomes of flowering plants of the USSR flora: Family Aceraceae - Menyanthaceae* (Nauka, Leningrad, 1990).
6. W. Poddubnaja-Arnoldi. Planta **19** , 46-86 (1933).
7. Ye. Ugolnikova, A. Kashin, A. Kondratyeva, Izv. Sarat. un-ta. Nov. ser. Chemistry, Biology, Ecology **17**, 53-61 (2017).
8. T. Leonova. *Chondrilla - Chondrilla L. In: Flora of the European part of the USSR* (Leningrad, 1989).
9. T. Kritskaya, A. Kashin, A. Parkhomenko, *GenBank: MH087455.1* (2018).

BIO Web of Conferences **11**, 00032 (2018) https://doi.org/10.1051/bioconf/20181100032
Prospects of Development and Challenges of Modern Botany

Soil and vegetal shifts in meadows under conditions of alluvial lithogenic matrix on islands in the delta of the Northern Dvina River

Tatyana Parinova [*], *Irina* Amosova, *and Elena* Nakvasina

Northern (Arctic) Federal University named after M.V. Lomonosov. Higher School of Natural Sciences and Technologies. Department of biology, ecology and biotechnology. Arkhangelsk, Russian Federation

Abstract. In this paper we represent the results of our research: a scheme of plant and soil succession changes on the alluvial lithogenic matrix in the zone of river and marine accumulation in the delta of the Northern Dvina river. We used more than 1000 geobotanical and 200 soil descriptions. The succession scheme of the transformation of meadow communities in the river accumulation zone is represented in this work. This scheme occurs both in the natural processes of phytocenosis and soil development, as well as in agrogenic influence, as well as a number of successional changes during formation of meadows in the zone of marine accumulation.

1 Introduction

In the European part of Russia, communities of the forest zone formed on the alluvial lithogenic matrix. Temperate climate causes an extensive development of mesophytic grass, shrub and tree vegetation. The delta of the Northern Dvina located on the border where the influence of the river itself meets the one of the White Sea. There is a multitude of islands in the delta where developed areas combine with natural ones. Natural and anthropogenic processes define a specific scheme of changes in soil-vegetation cover.

According to the botanical-geographical division of the Russian Federation, the research area is located within the Eurasian taiga (coniferous-forest) area. It has a zonal type of northern European spruce forests and intrazonal type of meadow vegetation.

Natural and climatic conditions in the area, combined, affect negatively on the distribution of meadow vegetation. But due to the specific ecological conditions in the delta, this influence partially overlaps, which makes it possible to spread primary and secondary meadows in high latitudes. Intensive use of meadows (as hayfields, pastures, arable land) and then a sharp decline in agriculture in the late XX – early XI century caused various changes in the ecological, floristic, productive, structural features of communities. Seaside meadows, as a rule, are not in use and are subject to natural changes.

Soil and vegetation cover of the islands in Northern Dvina delta has been studied fragmentarily. In the middle of the XX century a team of the Soil Science Department and

[*] Corresponding author: nadeinata@mail.ru

the soil and agronomic museum of the TSAA (now the Russian State Agrarian University named after K.A. Timiryazev) have carried out significant field work to survey the soils and vegetation cover of the meadows from the delta of the Vychegda River and to the sea islands in the delta [1, 2]. T.A. Parinova, E.N. Nakvasina, O.V. Sidorova provided recent ecology-floristic characteristics of flood meadows in the lower reaches of the Northern Dvina [3]. There are available point data for marshes and flood-meadows in Arkhangelsk region, mainly its mainland [4–6]. A group of researchers examined flood-meadows in the Murmansk region in detail [7–9].

2 Materials and Methods

Our research focused on two types of islands in the delta of the Northern Dvina in the Arkhangelsk region (North-West of the European part of the Russian Federation). T.A. Parinova and E.N. Nakvasina studied islands in the zone of river accumulation. (Islands: Andrianov, Lastolsky, Chubnavolskaya Koshka, etc.). E.Yu. Churakova and I.B. Amosova studied islands in the zone of marine accumulation (Golets, Kumbich). We investigated floristic composition and soils in those communities, using standard geobotanical methods. The majority of communities belong to the meadow type of vegetation. In total, we analyzed more than 1000 geobotanical descriptions, including 786 descriptions on trial plots of 1 m^2 and 250 descriptions on trial plots of 100 m^2. The names of the species of vascular plants are given in accordance with the summary of S.K. Cherepanov [10]. The names of mosses are in accordance with the list of mosses of Eastern Europe and Northern Asia [11]. We placed more than 200 basic (full-profile) soil sections and 100 extended soil sections. Classification and diagnostics of soils aligned with the classification of soils of 2004 [12].

3 Results

Based on the conducted studies, we constructed a general scheme of successive shifts in the conditions of the alluvial lithogenic matrix in the delta of the Northern Dvina River on the islands in the zone of river and marine accumulation (Figure 1). The fact that these shifts occur simultaneously with the combined effect of continental (zonal) conditions on the catchment area of the river and the erosion-accumulative and flood-alluvial activity of the river itself complicates the scheme above. This generally takes place in the context of the formation of the segmented floodplain of the Northern Dvina as a meandering river. Natural successional scheme violated through anthropogenic activities.

4 Discussion

In the zone of marine accumulation, seaside meadows emerge as primary meadow communities. These communities have limited distribution and small area, confined to the stony and sandy substrate of the islands in the White Sea and the loamy and clay substrates of the mainland. Seaside meadows are represented by mesophytic plant communities formed outside the tidal zone. Salty waters come here only during heavy storms. Meadows are located on the leeward sides of the islands – eastern (Mudyug Island) and south-western (Kumbish Island). This type of plant communities predominantly forms a narrow strip, with a width of 50 m, rarely extend to 100-200 m between marshes, and birch or pine forests. Seaside meadow communities are most well represented on the river alluvium of the maritime terrace of Kumbish island. They develop on peaty and sandy substrata. The total projective coverage is 40–55% (max = 90%). Such communities differ in relatively small height of the grass stand 30–40 cm. The species saturation in these communities is small, 15–20 species on average. There are no clearly expressed dominants in the grass stand, mainly the grasses predominate. Therefore, such communities are very bright and diverse. The most typical species are *Trifolium pratense* L., *Hylotelephium triphyllum* (Haw.) Holub, *Rumex acetosa* L., *Rumex crispus* L., *Oberna behen* (L.) Ikonn., *Dianthus superbus* L., *Stellaria graminea* L., *Tanacetum vulgare* L. and others. The grains are represented by: *Alopecurus arundinaceus* Poir., *Anthoxanthum odoratum* L. and several plants of the *Poa* species. The peculiarity of seaside meadows on this territory is the presence of species usually growing on salinized marshes (*Potentilla egedii* Wormsk., *Plantago maritima* L., *Puccinellia coarctata* Fern. & Weath.) and forest communities (*Avenella flexuosa* (L.) Drej.). The presence of halophytic species in the communities is due to the regular impact of sea water during particularly high tides and heavy storms.

The succession of vegetation within the sea and river terraces that have emerged from the main zone of the sea's influence is closely related to the relief conditions (elevation height above sea level), the nature of moistening, the degree of salinity of the substrate, and soil-forming processes. The climax and pre-climax vegetation is confined to absolute heights within the islands. Pioneer groups occupy silty drains, flooded with water at least twice a day. Plants do not form closed communities and are represented by single-or small-scale communities. Mostly in this zone grow such species as *Salicornia europaea* L., *Ruppia brachypus* J. Gay, *Puccinellia coarctata* Fernald & Weath., *Agrostis stolonifera* L., *Juncus filiformis* L. In some cases, thickets of *Phragmites australis* (Cav.) Trin d. ex Steud. and *Bolboschoenus maritimus* (L.) Palla may appear.

Marshing communities occur in small areas of 5–20 m from the water. Marshes are formed in a zone of periodic salinization, at least twice a year they are under water for a short time. Soil type – peaty or soddy silty substratum. This type of coastal communities is formed by halophytic and halomeophyte groups. Communities are small, the most common species of this zone are *Alopecurus arundinaceus*, *Atriplex nudicaulis* Bogusl., *Honckenya peploides* (L.) Ehrh., *Lathyrus maritimus* Bigel., *Leymus arenarius* (L.) Hochst., *Plantago maritime* L., *Triglochin maritimum* L., *T. palustre* L.

Seaside meadows are in the zone of episodic salinization – during severe storms, in 50–150 m from the water. Soils are sod-peaty. Formed mesophilic and halomezophilous plants, with some inclusion of halophytes (*Triglochin maritimum*, *Potentilla egedii*, etc.). In some cases, a narrow strip of meadow (not more than 50 m wide) is formed behind the meadows on a sandy substrate. Forest communities – birch forest or pine forests on dunes are formed in 150–200 m away from the water. Pine forests are located on the highest parts of the islands.

Most of the meadow communities in the zone of river accumulation in the delta of the Northern Dvina by their origin take an intermediate position between the primary and secondary community, since they have a dual exo-endodynamic genesis. Meadows appear as primary stages of vegetation, mainly on newly formed substrates (sand banks). Their existence is sustained as a result of anthropogenic-related activities (mainly grazing and mowing). The true secondary meadows that emerged from under the coniferous forest were not marked in the delta (in none of the soil profiles there are signs of podzol formation).

Transformation processes in soil and vegetation cover are interrelated. In the zone of river accumulation, soils are formed on modern alluvial deposits of different granulometric composition under the influence of three soil-forming macroprocesses: sod, meadow and marsh [13]. Due to the features of the slime, sandy or loamy soils predominate on various islands, often with visible signs of gleying. They are subject to regular flooding by flood waters and a deposition of fresh layers of alluvium on the soil surface. These processes determine their specific peculiarities of structure, water regime and genesis. Alluvial soils are "young", characterized by high biogenicity and intensity of soil formation, very diverse in regime, structure and properties.

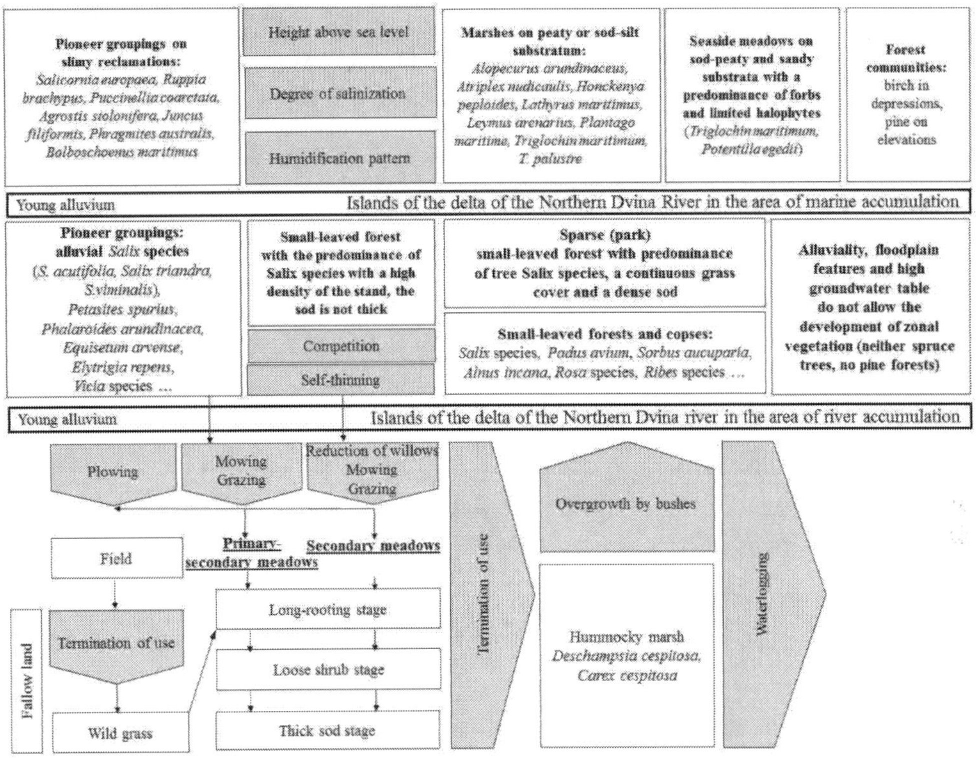

Fig. 1. Vegetative shifts on the alluvial lithogenic matrix in the zone of river and marine accumulation in the Deltaof the Northern Dvina River.

Without economic use under natural development, small pioneer communities in about 10 years are replaced by small-leaved forest with a predominance of *Salix* species with a high density of stand. As a result of competition and self-thinning, such communities pass to the stage of rarefied small-leaved forest with predominance of *Salix* species with tree-like forms, grassy cover with apparent sod. The herb layer is dominated by: *Elytrigia repens*, *Calamagrostis epigeios* (L.) Roth, *Veratrum lobelianum* Bernh., *Urtica dioica* L., *Filipendula ulmaria* (L.) Maxim., *Aegopodium podagraria* L., *Chrysosplenium alternifolium* L., *Thalictrum simplex* L., *Lysimachia vulgaris* L., *Vicia sepium* L. and sometimes *Matteuccia struthiopteris* (L.) Tod. Richness of species of herbaceous layer per 100 m^2 is 7–10 species. The tree layer is represented by: *Salix acutifolia* Willd., *S. caprea* L., *S. pentandra* L., *S. triandra* L., *S. viminalis* L. with an admixture of *Alnus incana* (L.) Moench, occasionally *A. glutinosa* (L.) Gaertn. The shrub layer: *Sorbus aucuparia* L., *Padus avium* Mill., *Ribes rubrum* L. and *R. nigrum* L. Moss layer: *Amblystegium serpens* (Hedw.) Bruch et al., *Plagiomnium ellipticum* (Brid.) T. J. Kop., *Brachythecium rivulare* Bruch et al., *Climacium dendroides* (Hedw.) F.Weber & D.Mohr. At this stage, the community comes to a dynamic equilibrium and can exist for many decades.

Economic activity violates the natural change of communities. In the past, alluvial lands in the delta were actively used as hayfields, pastures and arable land. Now they are almost all abandoned. Fallow communities in the delta experiencing regenerative shifts, which include the classical stages: weedy, rhizome, loamy and dense-bush.

Plowing and formation of fallow soils disrupt the natural course of soil changes. Self-regeneration of fallow soils follows the sod type. The analogue of the sod horizon – upper layer old plow horizon, densely intertwined with roots, begins to isolate itself in the first 10 years after the cessation of plowing. The signs of arable land in the upper horizons of soils are preserved up to 30–40 years and can be registered in deeper layers of the humus horizon compared to native soils, in the dark (black-brown) color of the mixed horizon and its even lower boundary. The disturbance of the activity of meliorative systems, widespread in the delta of the river. Northern Dvina, can lead to bogging flood-ground waters. In this case, gleying starts to develop in the lower horizons of the soil, and the erosion and erosion of the lower boundary of the humus (old–plow) horizon. As a result of self-restoration, arable soils undergo the stage of degraded soils (30–40 years), then they pass into subtypes of postagenogenic soils, the formation of which can follow an automorphic type (with the maintenance of the meliorative system) or hydromorphic (in violation of it), approaching the original (native) soils. There are probably two variants of the evolutionary development of soils during postagrogenesis and their transition to the native state: alluvial postragrogenic serohumus (peat) gley; alluvial postgragenic gray-humus.

Thus, the presented scheme reflects the main directions of soil-plant changes in the delta of the Northern Dvina. We have described four large shifts of natural origin in the marine accumulation zone and 3–4 natural shifts in the river accumulation zone. Anthropogenic transformation causes up to 9 different shifts.

This project is supported by grant from the Russian Foundation for Basic Research (RFBR) and the government of Arkhangelsk region № 17-44-290111 (project manager: Prof. Elena Nakvasina).

References

1. G.V. Afanasyev, Reports of the TSHA **3**, 259-264 (1957)
2. G.V. Afanasyev, Reports of the TSHA **34**, 193-198 (1958)
3. T.A. Parinova, E.N. Nakvasina, O.V. Sidorova, *Meadows of the lower reaches of the lower reaches of the Northern Dvina* (NArFU, 2013)
4. A.A. Korchagin, Scientific papers of Institute of the Academy of Sciences of the USSR **2**, 223-344 (1935)
5. V.M. Shmidt, V.G. Sergienko, Vestn. Leningr. un-ta **21**, 48-55 (1984)
6. E.Yu.Churakova, O.V. Sidorova, I.B. Amosova, Scientific notes PetrSU **2** 86-98 (2016)
7. N.V. Babina, Vegetation of Russia **3**, 3-21 (2002)
8. N.E. Koroleva, S.V. Chinenko, E.B. Sortland, Phyto diversity of Eas. Eur. **9**, 26-62 (2011)
9. L.A. Sergienko, *Composition and structure of vegetative cover of coastal ecosystems of the Russian Arctic* (Petrozav. state. un-t, 2012)
10. S.K. Cherepanov, *Vascular plants of Russia and neighboring countries* (Cambridge University Press, 1995)
11. M.S. Ignatov, O.M. Afonina, E.A. Ignatova, Arktoa J. **15**, 1-130 (2006)
12. L.L. Shishov, V.D. Tonkonogov, I.I. Lebedeva, *Classification and diagnostics of soils in Russia* (Ojkumena, 2004)
13. G.V. Dobrovolsky, *Soils of the river floodplains of the center of the Russian Plain* (MSU, 2005)

The dynamics of free amino acids in the grains *Tripogon chinensis* (Franch.) Hack. when germinating in the conditions of the osmotic stress

Irina Plyaskina[1,*], *Evgenii* Bondarevich[1], *Igor* Boriskin[2], *Natalia* Kotsyurzhinskaya[1] and *Ludmila* Ishina[2]

[1]FSBEI HE "Chita State Medical Academy", Chita, 672000, Russia
[2]Transbaikalian Agrarian Institute – branch of FSBEI HE "Irkutsk State Agrarian University named after A.A. Ezhevsky ", Chita, 672023, Russia

Abstract. The content of free amino acids in seeds and sprouts of *T. chinensis* in the control and at an osmotic pressure of 5 atm. was determined. The group of acidic and basic amino acids predominates in seeds. An adaptive reaction to a physiological drought is the transformation of the free amino acids metabolism. This is expressed in an increase in the total amount of free amino acids at the osmotic pressure of 5 atm., providing the osmotic component of adaptation. Under the conditions of the osmotic stress, the concentration of acidic and basic amino acids increases up to 48 hours, this may be due to the continued hydrolysis of reserve proteins. Under the influence of the osmotic stress there are changes in the group of amino acids, the metabolic precursor of which is glutamic acid. The amino acids, a part of this group, exhibit protective, signaling properties. Thus, the increase in the arginine concentration and ornithine is noted; this indicates the activation of the ornithine cycle and on the increase of amino acids catabolism. The revealed features can ensure the successful germination of *T. chinensis* grains under the conditions of the physiological drought.

1 Introduction

One of the mechanisms ensuring resistance to the stresses of different nature (cold, drought, hypoxia) is the accumulation of free amino acids [1-4]. All types of influences are reflected at the level of free amino acids. Cold and drought cause an increase in the amount of free amino acids, and hypoxia leads to a decline in their concentration. The glutamic acid and

* Corresponding author: thebestdamnthing@mail.ru

substances, for which it is a metabolic predecessor (glutamine, proline, ornithine, arginine, γ-aminobutyric acid) dominate in stress conditions. From this group, the functions of γ-aminobutyric acid and proline are most widely discussed as stress metabolites [5]. The intensity of accumulation of alanine, proline, hydroxyproline and glycine in deficit of water can change due to the use of these amino acids on the synthesis of the corresponding betaines, known for its osmoregulation properties [6]. The great part of research works on studying of amino acid composition are related to species of nutritional, fodder or pharmacological value [7, 8]. Most often such cereals as wheat, corn, rice, barley, sorghum, triticale are the objects of these researches [9, 10]. The wild cereals are insufficiently studied in this aspect. On the territory of the Eastern Transbaikaliye, the plants of a xeromorphic nature are prevalent; they are resistant to the conditions of habitat with a lack of moisture. One of such kind of plants is *Tripogon chinensis* (Franch.) Hack. – a relict cereal which is growing on the petrophytic plots of the slopes where it is exposed to extreme environmental factors (temperature fluctuations, insufficient moisture, insolation, etc.). Therefore, the study of the effect of conditions that imitate a physiological drought on the dynamics of free amino acids can contribute to the definition of mechanisms for the adaptation of the wild cereals of the Eastern Transbaikaliye. The aim of the study is to determine the effect of osmotic stress on the dynamics of free amino acids in seeds and sprouts of *T. chinensis*

2 Materials and Methods

2.1 Object of study

T. chinensis is the only representative of the genus *Tripogon* in Russia, although this tropical genus noted in Asia and Africa, includes about 30 species, 1 species in America [11]. The species lives in strictly defined biotopes, growing in the petrophytic steppes along stony tops of hills, on the surfaces of rocks and large boulders, often forming a three-beard low-grass steppe [11-13]. This drought-resistant mountain-steppe species of eastern Asia is located in the Eastern Transbaikalye on the northern border of its natural habitat. The habitat of the species covers the southern regions of the Transbaikalian Region, Manchuria, Korea, Mongolia and is mentioned in the Jewish Autonomous Region and in the Khabarovsk Territory.

Only seed reproduction is characterized for the *T. chinensis*. Therefore, the adaptations that ensure the germination of seeds and the further growth of *T. chinensis* in extreme environmental conditions are of particular interest. The material for the study was mature purified *T. chinensis* grains harvested in 2014 in the Akshinsky district of the Transbaikalian region and sprouts (24, 48 and 72 hours).

2.2 Methods

The germination of seeds was carried out in the dark at 20 ° C during three days. The osmotic stress was created with an aqueous solution of mannit with a concentration of 0.208 mol / l, which corresponds to an osmotic pressure of 5 atm. Control groups of grains were germinated on filter paper moistened with distilled water, and experienced – in the presence of osmotics, which led to the formation of a physiological drought. Every day the amount of germinating seeds was determined (in %). For the determination of free aminoacids, weights of cereal grains (dry and germinated) weighing 50 ± 0.2 mg were used. The extraction of free aminoacids was carried out with a 0.89 % solution of sodium chloride. The water-salt extract was dissolved in acetonitrile (1:1). The content of free aminoacids was determined by the method of the high-effective liquid chromatography [14]. The content of 20 aminoacids is determined, 2 of which are non-proteinogenic (taurine and ornithine). The obtained data was subjected to statistical processing with the help of the package Microsoft Excel 2010 and PAST 3.0 [15].

3 Results and Discussion

The visible seed germination in the control and at an osmotic pressure of 5 atm. was observed on the second day, but under the osmotic stress conditions the germination rate was lower (71.1 and 45.1 % accordingly). On the third day the germination rate in all samples was 89 %. The dynamics of free amino acids was studied, See Table 1.

Table 1. Dynamics of free amino acids (in μg / g in wet mass) in seeds and sprouts of *T. chinensis* under the conditions of the osmotic stress

Amino acids	Dry seeds	Control			5 atm.		
		24h	48h	72h	24h	48h	72h
Asp	13,9	62,15	25,38	300,94	62,74	261,77	140,86
Glu	47,2	216,76	88,52	625,62	213,22	933,31	443,47
Asn	51,51	180,9	73,87	1252,42	232,77	601,44	343,95
Gln	88,78	499,84	204,12	2385,08	401,21	537,67	1243,13
Ser	9,38	229,27	93,63	290,73	42,4	787,5	125,77
Gly	33,03	187,97	76,76	612,94	149,28	714,37	257,4
Thr	52,39	64,79	26,46	99,78	236,76	183,68	49,04
Ala	5,37	230,03	93,94	135,14	24,25	845,54	5,9
His	285,04	94,45	38,57	286,19	499,18	377,32	144,76
Lys	31	38,67	15,79	68,91	140,08	171,47	25,67
Arg	11,49	50,03	20,43	36.14	51,93	187,1	20,95
Trp	12,6	48,16	19,67	150,39	56,94	182,89	41,56
Phe	32,34	66,95	27,34	229,05	146,14	168,89	109,09
Tyr	16,52	94,43	38,56	104,74	74,63	229,16	40,68
Met	8,14	21,82	8,91	74,36	36,77	31,95	23,74
Val	26,64	144,04	58,82	188,01	115,86	430,86	70,11
Ile	18,35	64,26	26,24	129,01	82,93	236,35	54,28
Leu	14,54	125,98	51,45	157,48	65,73	412,56	74,16
Taurine	7,72	139,54	56,99	21,31	34,89	1088,4	32,07
Ornithine	6,64	23,27	9,5	11,76	30,02	54,45	37,79
Amount of free amino acids	772,58	2583,31	1054,95	7123,86	2697,73	8436,68	3284,38

In the dry grains of *T. chinensis* the group of acidic and basic amino acids predominates, what is characteristic as well of seeds of grain cereals [3, 16]. Especially the histidine content is great (285.04 μg / g in wet mass). According to the dynamics of the number of acidic amino acids and their amides the following features were noted. In comparison with dry grains in the first 24 hours from the onset of the germination the concentration of Asp and Glu was increased more than in 4 times, and of amides – more than in 5 times. The similar results were observed under conditions with an osmotic pressure of 5 atm. At 48 hours from the onset of germination in the control the number of all the amino acids in the group decreased and increased again to 72 hours. The contents of Gln and Asn, the main donors of the amino groups, had the maximum values. Under conditions of osmotic stress the maximum for Glu was noted at the time of 48 hours, and the maximum for Gln after 72 hours.

On the first day there was an increase in the number of Ser and Gly, to the second day in the control – a significant portion of these amino acids are consumed, and after 72 hours their concentration increases again. Whereas under the conditions of osmotic stress the amount of Ser and Gly continued to increase up to 48 hours, reaching the maximum values (787.5 and 714.37 μg / g in wet mass) and the decrease occurred at 72 hours. A similar trend is noted for Ala. The concentration of threonine in the control in comparison with dry grains changed insignificantly, whereas under stress conditions the increase in its content was significantly greater. The main amino acids were characterized by the following dynamics features. In control the consumption of His lasts during the first day and under the osmotic stress the concentration of this amino acid increases almost in 2 times. Also, osmotic stress leads to an increase in the concentration of Lys and Arg up to 48 h, but then their concentrations were sharply decreased. A number of researches approves increase of quantity of arginine in a drought [17, 18]. At 5 atm. on second day concentration of arginine was maximal (187,1 μg / g in wet mass), further concentration decreases, that coincides with activization of growth processes. Amino acids of ornithine cycle (arginine, ornithine, citrulline) accumulate the amine nitrogen which is released hydrolysis of proteins in stress conditions, are metabolic predecessors of polyamines performing protective functions [19]. It can be assumed that the osmotic stress is carried more successful by this type of cereals in connection with the dynamics of ornithine. This metabolite reflects the level of the pathway work for free ammonia binding.

The aromatic and hydrophobic neutral amino acids are emitted in the first days as a result of a proteolysis of the reserve proteins and this process at 5 atm. occurs up to 48 h. This can explain a large amount of the free amino acids at 5 atm. (8436,68 in μg / g in wet mass).

Thus, the osmotic stress leads to a slowdown of the water flow into the seed, this is reflected on the processes that ensure the transition to the start-up of growth. The study of the dynamics of free amino acids in *T. chinensis* showed that during the swelling process an increase in the concentration of all amino acids (except His) occurs due to the hydrolysis of reserve proteins. The growth processes, actively coming from the second day, lead to the involvement of amino acids in the synthesis of metabolites, this is accompanied by a drop in the concentration of amino acids.

References

1. O.I. Yakhin, A.A. Lubyanov, Z.F. Kalimullina, R.A. Batraev, Izvestia Orenburg state agrarian univ. **33**, 38-40 (2012)
2. N.E. Sudachkova, I.L. Milyutina, L.I. Romanova Free, J. Stress Physiol. **3**, 4-14 (2007)
3. S. Movafegh, R. Jadid, S. Kiabi, Jour. of Stress Phys. Biochem. **8**, 157-168 (2012)

4. E.V. Alaudinova, P.V. Mironov, Khimija Rastitel'nogo Syr'ja **3**, 85–91 (2017)
5. A.W. Bown, B.J. Shelp, Plant Phys. **115**, 1-5 (1997)
6. A.D. Hadson, B. Rathinasabapathi, J. Rivoal, M. Burnet, M.O. Dillon, D.A. Gage, Plant Biol. **91**, 306-310 (1994)
7. B. Pisarikova, S. Kracmar, I. Herzig, Czech J. Anim. Sci. **50**, 169–174 (2005)
8. J.H. Umrani, V.M. Pahoja, I.T. Ansari, M.A. Sahito, B.U. Panhwar, S.P. Tunio, Journ. of Agriculture and Veterinary Scien. **2**, 18-20 (2013)
9. A.C. Stringfellow, J.S. Wall, G.L. Donaldson, R.A., Cereal Chem. **53**, 51-60 (1975)
10. J. Mosse, J.C. Huet, J. Baudet, Cereal Chem. **65**, 271-277 (1988)
11. N.N. Tzvelev, *The cereals of USSA* (Nauka, Leningrad, 1976)
12. M.N. Lomonosova, *Flora of Siberia.* Vol. 2. *Poaceae (Gramineae)* (Nauka, Novosibirsk, 1990)
13. N.S. Probatova, *The Red Book of Russian Federation (plants and mushrooms)* (KMK Scientific Press Ltd., Moscow, 2008)
14. T. Teerlink, Clinical Chem. **42**, 245-249 (1994)
15. Ø. Hammer, D.A.T. Harpe, P.D. Ryan, Palaeontologia Electronica **4**, 1-9 (2001)
16. P.E. Padgett, R.T. Leonard, Journ. of Exp. Botany **47**, 871-883 (1996)
17. S. Kawasaki, C. Miyake, T. Kouchi, A. Yokota, Plant Cell Phys. **41**, 864–873 (2000)
18. P. Diaz, O. Borsani, A. Marquez, J. Monza, Lotus Newsletter **356**, 83-92 (2005)
19. J.H. Liu, H. Kitashiba, J. Wang, Y. Ban, T. Moriguchi, Plant Biotechnol. **24**, 117-126 (2007)

Cone quality and seed efficiency in the clones from mutational witches brooms of *Pinus sibirica*

Olga Polyakova[1,2,*], *Evgeniya* Zhuk[1], and *Sergei* Goroshkevich[1]

[1]Institute of Monitoring of Climatic and Ecological Systems, Russian Academy of Sciences, Siberian Branch 634055 Tomsk, 10/3 Academichesky ave., Russia
[2]Tomsk State University, 634050 Tomsk, 36 Lenina ave., Russia

Abstract. Mutational witches' broom is a fragment of the tree crown with abnormal morphogenesis including slow shoot growth, abundant branching and decreased apical dominance which caused by somatic mutation that presumably originates in the apical meristem bud. To study the variability in the cone quality and seed efficiency in witches' broom clones of *Pinus sibirica*, the cone size and structure were assessed in 21 mutant clones. The mutant clones were significantly differentiated both by the cone traits and seed quality. Their cone size varied from less than 3 to 4 cm. The clones had large number of sterile scales in proximal and distal cone parts, and from 16 to 39 fertile scales per cone. Cones from all mutant clones were inferior to average normal cones of the species. Six clones had large portion of filled seeds, more than 70%, and the other clones had less than 50% of filled seeds. The clones had 10-33% of empty seeds and 1-73% of seeds with collapsed megagametophyte per cone. In general, the seed quality in mutants was quite low, but individual clones had almost normal seed quality, and their seed efficiency was within the species norm.

1 Introduction

Mutational witches' broom (WB) is a fragment of the tree crown with abnormal morphogenesis including slow shoot growth, abundant branching and decreased apical dominance A somatic mutation that presumably originates in the bud in the apical meristem is considered as a cause of the WB [1]. Unlike pathological WB, mutational WB are sporadically distributed and have a healthy appearance, normal reproduction, and normal longevity [2-3]. Mutational WB is a major source of the majority of ornamental dwarf conifer cultivars [4]. To date, certain patterns in the morphological traits variability of WB clones were revealed [5]. Although the seed progeny involvement in breeding process has great importance, the information concerning the variability of cone structure and seed efficiency in mutant clones is scarce. The objective of the work was to study the variability of the cone quality and seed efficiency in WB clones of *Pinus sibirica*.

* Corresponding author: polyakova_olga93@mail.ru

2 Materials and Methods

2.1 Plant material

The investigation was carried out on grafted WB clones of *Pinus sibirica*. The experiment was conducted at the Kedr Field Station, which is managed by the Institute of Monitoring of Climatic and Ecological Systems, Russian Academy of Sciences, Siberian Branch (56°13'N 84°51'E, 78 m above sea level, south-east of West Siberian Plain, southern taiga). Plant material for grafting were 21 trees with morphologically different mutational WB. Parental WB had different crown density, which was defined by the combination of three main traits, abundance of branching, shoot length and degree of apical dominance. The more abundant branching, shorter shoots and lower apical dominance, the WB looks denser.

In 2007, scions from each WB and NC were grafted onto local five-year-old Siberian stone pine rootstocks. The grafted plants were grown with 1×0.5 m spacing in rows, each clone contained 10-20 ramets. At the clone age of 10 years, cones from all cone-bearing ramets were collected. If a clone had weak cone-bearing, a whole cone crop (3-9 cones) was taken for analysis. If a clone had abundant cone-bearing, 15 cones from the whole crop were randomly taken for analysis.

2.2 Measurements and data analysis

The following traits were measured in the cones: cone length, cone diameter, number of sterile scales in proximal cone part, number of fertile and sterile scales in medial cone part, number of sterile scales in distal cone part, number of rudimentary and full-grown seeds. Number of unpollinated ovules was calculated as doubled number of sterile scales in medial cone part. Among full-grown seeds the number of filled and empty seeds, and also number seeds with collapsed megagametophyte were determined using X-ray analysis. Seed efficiency was calculated as portion of filled seeds from the total number of ovules. Normality was tested using the Kolmogorov-Smirnov test. If the distribution was normal, ANOVA and Newman–Keuls test were used to compare clones. In the other cases, the Kruskal-Wallis and non-parametric Dunn's multiple comparison tests were used.

3 Results

All cone traits were highly variable in WB clones. Fifteen percent of clones had very small cones with length and diameter less than 3 cm. Twenty percent of clones had quite big cones compared with other clones, their length were more than 4 cm and diameter were more than 4.5 cm. The other clones had intermediate cone size. All the clones had large number of sterile scales in proximal and distal cone parts (Fig. 1). In some clones more than half of total number of cone scales were sterile scales. The number of fertile scales in medial cone part was even more differentiated among clones than cone size, from 16 to 39 fertile scales per cone.

All the clones had almost equally high number of unpollinated ovules in medial cone part, about 25% (Fig. 2). In general, number of rudimentary seeds was quite low with some variations among the clones. Number of full-grown seeds varied significantly among the clones, from 6-11 to 36-42 full-grown seeds per cone. Seed efficiency was from 5 to 48% in different clones and 27% in average.

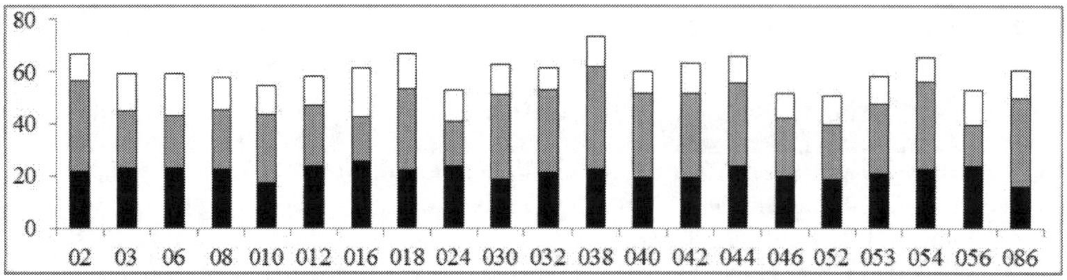

Fig. 1. Number of cone scales in mutant clones: sterile scales in proximal cone part (black), fertile scales in medial cone part (gray), sterile scales in distal cone part (white)

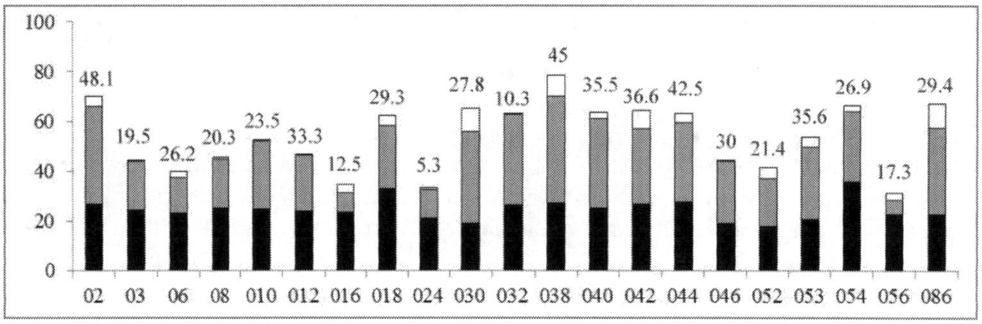

Fig. 2. Number of different types of seeds in mutant clones: unpollinated ovules (black), full-grown seeds (gray), rudimentary seeds (white). The values above the columns are seed efficiency (%)

The quality of full-grown seeds also significantly varied among the clones. Six clones had large portion of filled seeds (more than 70%), half of clones had medium portion of filled seeds (about 50%), and the other clones had low portion of filled seeds (less than 40%) (Fig. 3). All the clones had about 20% of empty seeds, except 2 clones having much more empty seeds. Portion of seeds with collapsed megagametophyte significantly varied among the clones. In more than half of clones their portion was moderate, but several clones had 30-70% of these seeds.

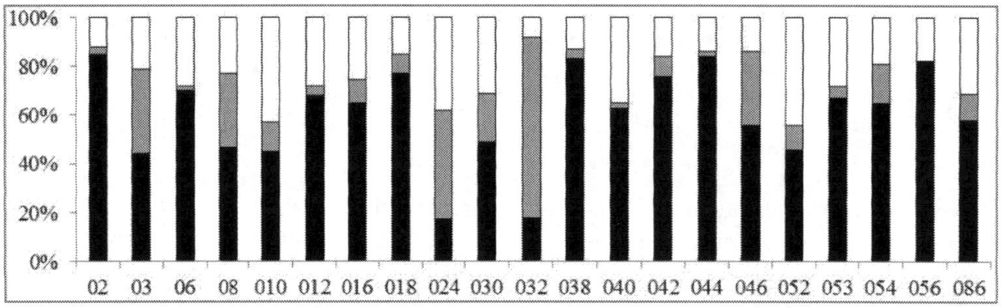

Fig. 3. Quality of full-grown seeds in mutant clones: portion of filled seeds (black), seeds with collapsed megagametophyte (gray) and empty seeds (white)

4 Discussion

Mutational WB cones are generally smaller than normal cones in some species. This pattern is typical both for soft pines, such as *P. strobus* [3], and for hard pines, such as *P. sylvestris* [6], *P. banksiana* [7] and *P. halepensis* [8]. Usually WB cones were 1.2-2-fold smaller, however, in some cases WB cones did not differ from normal cones [9]. Average cone length in *P. sibirica* growing in wild forest is 7-8 cm [10]. The mutant clones had 2-fold smaller cone size, and there were no clones with cone size at least approximately close to normal species cones.

Number of cone scales is closely related to the cone size. The mutant clones with the smallest cones had the smallest number of scales. It is possible that the cone size and the specific cone structure with large portion of sterile scales are related to the specific WB crown structure [5]. Large portion of unpollinated ovules could be caused by abnormal density of WB crown and associated pollination problems.

Quality of full-grown seeds were not related to the cone size. A clone with typical WB cones, that were good enough visually, had largest portion of seeds with collapsed megagametophyte. Among *Pinus sibirica* trees there are also ones producing cones with low quality seeds [11], therefore, it was impossible to determine whether low seed quality is WB mutations attribute or it depend on parental tree features. In general, WB clones of had low seed quality, however, individual clones had almost normal seed quality, and their seed efficiency was within the species norm.

This work is supported by the Russian Science Foundation (grant no. 18-16-00058).

References

1. J. Duffield, J. Wheat, Silvae Genet **12**, 129-133 (1963)
2. D.C. Buckland, J. Kuijt, For. Sci. **3**, 236-242 (1957)
3. A.J. Fordham, Arnoldia **24**, 29-50 (1967)
4. A. Farjon, *A handbook of the world's conifer* (Brill, Leiden-Boston, Netherlands, 2017)
5. E. Zhuk, G. Vasilyeva, S. Goroshkevich, Trees **29**, 107+-1090 (2015)
6. V.V. Shul'ga, Lesovedenie **3**, 82-86 (1979)
7. T.D. Rudolph, D.M. O'Malley, E.A. Reed, *Proceedings of the North Central Tree Improvement Conference*, 162-174 (1983)
8. P. Vrgoc, *Proceedings of XX International Eucarpia Symposium, Section Ornamentals, Strategies for New Ornamentals*, 199-205 (2002)
9. V.I. Noskov, S.F. Negrutzkiy, Sci. Notes of Voronezh For. Engineering Institute **15**, 207-211 (1956)
10. V.N. Vorobyev, *Biological basis of complex use of stone pine forests* (Novosibirsk, Nauka, 1983)
11. E.V. Titov, *Kedr. The king of the Siberian taiga* (Moscow, Kolos, 2007)

Age structure of coenopopulation of *Fragaria vesca* L. in the natural and transformed habitats

Natalya Romanova[1*], and *Irina* Olefirenko[1]

[1]Kemerovo State University, Kemerovo, 650000, Russia

Abstract. The article describes the types of ontogenetic spectra of *Fragaria vesca* L. growing in the natural and broken habitats. It shows that the character of age spectra did not differ and invariably remained in both ecotopes for the years of the research. The old individuals were frequently found in an atypical habitat than in natural one. The nature of self-maintenance of coenopopulation of a species is typical for its biomorph and does not depend on the place of growth.

1 Introduction

The populational and ontogenetic method of researches gained wide popularity because the various indicators which characterize development and the state of a species are considered with its help in the different phytocenotic conditions [1].

The purpose of this work is to reveal the features of age structure of coenopopulation of Fragaria vesca L. in the natural and broken habitats.

2 Materials and methods

Fragaria vesca L. is a perennial, polycarpic, ground-stoloniferous plant of the Rosaceae family [2]. The wild strawberry is an ecologically plastic species [3, 4]. It actively acclimatizes anthropogenically broken habitats revealing itself as a typical explerent [5].

The researches were conducted near the township Privolniy in Kemerovo district in 2015–2017 in the first decade of July. Two coenopopulations of wild strawberry were chosen: CP1 is a control one (natural typical habitat) located in the birch forest, rich in herbs and cereal, and occupied the area of 2–2.5 thousand m^2, and CP2 is an experimental one (transformed atypical habitat) on the slope of the southern aspect of the railway dump of rock with the area of 4 thousand m2 with forming cereal and rich in herbs phytocenosis. The distance between the studied coenopopulations is 700 m [6].

One counted the vegetative and generative rosettes on the sampling quadrats of 1 m^2. There were 30 sampling quadrats grounded in the forest and on the dump. The age conditions of individuals were defined in coenopopulations and the age spectra were interpreted on S.A. Dubrovnaya [4]. The age spectra were assembled with the use of the

* Corresponding author: chatn@yandex.ru

standard populational and ontogenetic techniques for the definition of ratios of different age groups in coenopopulations.

3 Results

The analysis of ontogenetic spectra of 2015 showed that coenopopulations were normal incomplete during the research. This feature is noted for the wild strawberry growing in the broken and undisturbed habitats of the European part of the areal [4].

CP1 was characterized by a left-sided age spectrum with the absolute dominance of virginile individuals (68%) (fig. 1a). The obtained spectrum is similar to S.A. Dubrovnaya's one [4] for coenopopulations of wild strawberry growing in the conditions of low illumination.

CP2 was characterized by a left-sided age spectrum (45% of individuals of pre-generative fraction) (fig. 1b). At the same time the part of middle-aged generative individuals made 28% and the part of post-generative individuals was less than 1%.

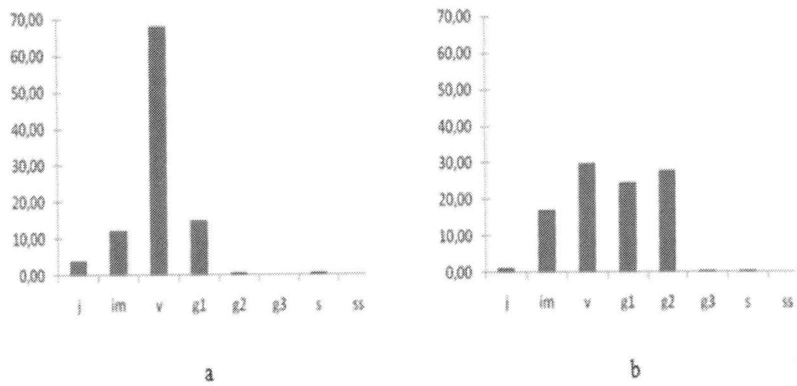

Fig. 1. Age structure of *F. vesca* in CP1 (a) and CP2 (b), 2015. Notes: x-direction – age conditions, y-direction – part of individuals, %.

The coenopopulations were also normal incomplete in 2016. There were no senile individuals in the age spectra of wild strawberry on the dump and the individuals of old generative, sub-senile and senile age were absent in the forest.

The age spectrum of CP1 was characterized as left-sided one with the greatest parts of individuals of the pre-generative period (immature and virginile). The part of middle-aged generative individuals made 25% (fig. 2a). The old individuals were absent.

The age spectrum of CP2 was characterized as left-sided with the dominance of immature, virginile and young generative rosettes (fig. 2b). In the work by L.B. Zaugolnova (1994) and S. A. Dubrovnaya [4], the spectra like those obtained by us for the control are called bimodal. The part of middle-aged generative individuals made 20% and about 8% of post-generative ones.

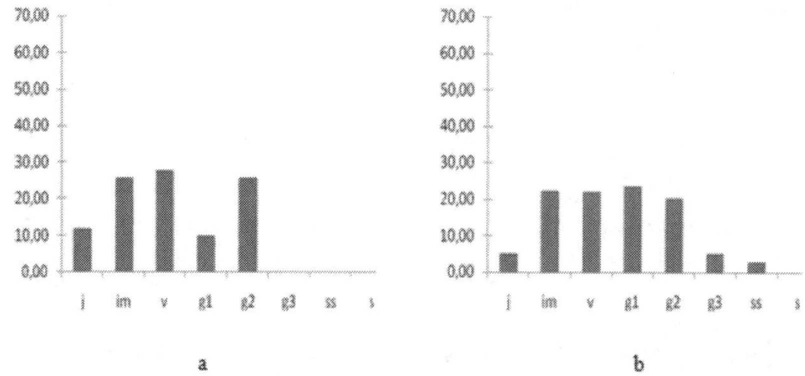

Fig. 1. Age structure of *F. vesca* in CP1 (a) and CP2 (b), 2016. The same notes as fig.1.

In the analysis of ontogenetic spectra of 2017, we revealed that coenopopulations were normal incomplete as well as in previous years. There were no senile individuals in transformed habitat and there were no old generative and senile ones in natural habitat.

CP1 was characterized by a left-sided age spectrum with high content of immature (28%) and virginile (41%) individuals. The part of generative individuals made 13% and the part of post-generative individuals was 15% (fig. 3a).

The age spectrum of CP2 was also left-sided (51% of individuals of pre-generative fraction) (fig. 3b). At the same time the part of generative individuals made 48% and the part of post-generative individuals was about 1%.

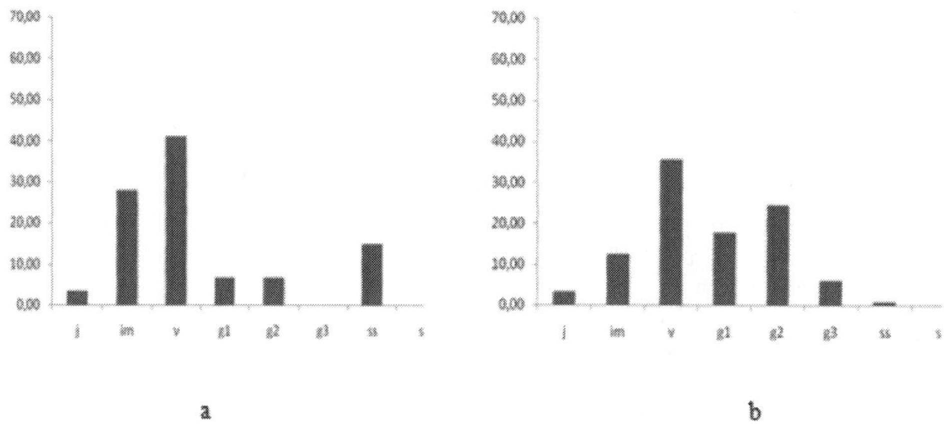

Fig. 3. Age structure of *F. vesca* in CP1 (a) and CP2 (b), 2016. The same notes as fig.1.

4 Discussion

The coenopopulations of *F. vesca* of various ecotopes were characterized as normal incomplete that corresponds to the literature data. The type of ontogenetic spectrum of wild strawberry in control and experimental coenopopulations was left-sided and did not change by years and by habitat type as well. Our results do not altogether agree with the data of other authors. For example, S.A. Dubrovnaya [4] obtained a left-sided spectrum for coenopopulations of *F. vesca* growing in the European part in the conditions of low illumination and she obtained bimodal spectrum in the open habitats. However, T.P. Larkina has the results corresponding to ours after conducting similar researches in the South Ural [7].

At the same time in days of the research, the ratio of rosettes of various age conditions varied in the spectra. In certain years, the dominance of virginile and generative individuals was noted in the natural and transformed habitats. The juvenile individuals were very few in both coenopopulations (2–12%) as well as the old ones (0–15%). It should be noted that old individuals are often found in the coenopopulation on a dump slope.

The character of age spectra of the studied coenopopulations of *F. vesca* indicates the mixed type of their self-maintenance with the dominance of vegetative one that corresponds to a vital form of the species.

References

1. V.N. Ilyina, Samarian Luka: Reg. and Glob. Ecol. Probl., **19**, 99 (2010)
2. *Ontogenetic atlas of medicinal plants* (MarSU, Yoshkar-Ola, 1997)
3. L.G. Dobrenkova, *Physiological features of strawberry resistance to drought and salinity* (LGU, Leningrad, 1978)
4. S.A. Dubrovnaya, *Structure of natural wild strawberry populations: Fragaria vesca L.* (MGU, Yoshkar-Ola, 2000)
5. J. Philip Grime, S. Pierce, *The Evolutionary Strategies that Shape Ecosystems* (Wiley-Blackwell, Chichester, 2012)
6. N.G. Romanova, *Problems of botany of south Siberia and Mongolia* (AGU, Barnaul, 2015)
7. L.B. Zaugolnova, Struct. of Seed Plant Pop. and Probl. of Their Monit., 70 (1994)
8. T.P. Larkina, Perm. Agr. Rep. **2**, 116 (2017)
9. J. Schulze, R. Rufener, A. Erhardt, P. Stoll, PE **54**, 369 (2012)

Species composition dynamics in successive plant assemblages on the northern slopes of the coal mining spoils in the arid areas of Khakassia

Olga Safronova[1], *Tatiana* Lamanova[2], *Natalia* Sheremet[2,*], and *Vladimir* Doronkin[2]

[1]Research Institute of Agricultural Problems of Khakassia, 655132 Abakan, Russia
[2]Central Siberian Botanical Garden of SB RAS, 630090 Novosibirsk. Russia

Abstract. The results of the study of the species composition dynamics in successive plant communities developing on the northern slopes of the coal mining spoils dating back to the 70s, 80s, 90s and 2000s years in the arid areas of Khakassia, are presented for the first time. During the first 10-15 years of revegetation there was an increase in taxonomic diversity in the successive plant assemblages. Endemic species of the *Fabaceae* family, common for the bunch grasses steppes and stony steppes of Khakassia, were not found on the northern spoil slopes. Mesoxerophytes were the most abundant ecological group. Drastic increase in the complexity of the successive plant assemblages in the arid areas of Khakassia was found to take place after 25-30 years of revegetation.

1 Introduction

Due to the surface coal mining huge areas become severely disturbed by such technogenic impact. Decreasing the negative effect is possible by restoring vegetation cover and biological productivity of the disturbed territories. Development of plant cover during revegetation of the coal mining soils has been receiving much attention of researchers in many countries [1, 2, 3, 4].

Coal mines in Khakassia are located along the routes of inverse aerial mass movement, bringing with them into the Chernogorsk and Abakan hollows, as well as other areas, plenty of dust from the explosive surface openings. In winter time with few winds all the industrial emissions stay above the densely populated cities and towns, slowly falling to the surface and hence negatively affecting human health. Plant cover that develops during revegetation of the spoils somewhat mitigates such negative effects.

The aim of the research was to study floristic composition of plant assemblages developed due to the revegetation succession on the coal mining spoils.

* Corresponding author: nsheremet@yandex.ru

2 Methods and environmental conditions

The objects of the study were ecosystems located on the northern slopes (with the most favourable conditions) of the coal mining spoils formed by the Chernogorsk mine (LLC "SYEK-Khakassia") in the 70s, 80s, 90s and 2000s years, i.e. at 10 years increment, in the process of open mining.

Since 2008-2011 aboveground plant parts of different species were clipped to assess aboveground phytomass of a species and determine species composition of the economically significant botanical groups growing on the spoils. To study the vertical structure of aboveground phytomass the standing parts were cut down to the roots, then cut in 0–3, 10-20 and 20-30 cm layers and weighed as fresh mass [5]. Geobotanical description of the plant assemblages was carried out according to A.G. Voronov [6]. Projective plant cover (percentage) was assessed visually, while species abundance was determined according to the Drude scale. The phytocenosis type was determined according to the set of dominating species. Belowground phytomass storage and its vertical layer distribution was done by sampling soil monoliths in triplicates from the surface area of 10 dm3 and 0-10 and 10-20 cm layers only, as sampling from the deeper layers was not feasible. Roots were washed in water, then dried and weighed.

The climate of the region is sharply continental with cold long winters and hot dry summers. Average yearly temperature ranges +1.8 to –1.0 ^{0}C. The absolute temperature minimum occurs in January (–40.6 ^{0}C), while the average maximum occurs in June +35.6 ^{0}C.

According to the geobotanical zoning by Kuminova et al. [7], the area of the Chernogorsk coal mining spoils can be regarded as part of the Near-Abakan (Central Khakassia) district of the Minusinsk hollow, where the most typical are the bunch grass and true steppes in a typical version of four-grasses steppe, distinguished by V.V. Reverdatto.

3 Results and discussion

Floristic composition of plant assemblages on the northern slopes of the technogenic chronosequence of coal mining spoils was found to consist of 40 species of higher vascular plants, representing 35 genera and 18 families. The most species-rich families were *Asteraceae* and *Poaceae*, accounting for 30% and 22.5%, respectively. Prevalence of these families is indicative of the steppe type of vegetation succession.

The number of species representing the *Fabaceae* family is the highest among other families comprising the flora of the undisturbed regional bunch grass [7] and stony steppes of Khakassia. It is related to the fact that Minusinsk-Khakssia steppes since the end of Pleiocene have been the centre of the autochtonous development of legumes, which is confirmed by significant endemism of the regional flora [9]. However, endemic species, representing *Fabaceae* family, were not found on the studied coal mining spoils even after 40 years of revegetation.

Among multi-species genera found in stony and small tussock steppe were *Astragalus*, *Oxytropis*, *Artemisia* and *Potentilla*. According to L.I. Malyshev [10], increasing floristic importance of these genera is most likely due to the continental climate and montane relief.

On the northern slopes of the coal mining spoil chronosequence these multi-species genera were represented by two genera, namely *Artemisia* with 5 species and *Cirsium* with 2 species.

The number of plant species, families and genera were found to increase significantly over the first 20 years of the spoil revegetation (Tab.).

Table. Dynamics of the species composition of plant assemblages developed on the northern slopes of the coal mining spoils in the arid area of Khakassia

Floristic composition	The time of spoil formation			
	2000s	90s	80s	70s
Families	5	12	12	14
Genera	9	24	21	29
Species	9	29	24	34

All the observed species belonged to three ecological groups: the most represented were mesoxerophytes, contributing 41-46% into the standing aboveground phytomass, the second largest group were xerophytes, contributing 21-33%, followed by mesophytes with their 22-28%. In the stony steppes of Khakassia xerophytes account for 22% of the total number of species, while mesoxerophytes contribute 34%; in the flora of bunch grass steppes xerophytes and mesophytes account for 49% of the total number of plant species.

Species richness can be considered as an indicator of a better or worse ecosystem status. Theoretically there is a transition from the geometric range via log-range and log-normal distribution towards the "broken shaft" model. Such transition follows increasing complexity of communities. Geometric range can be observed in case of very few species dominance with very low abundance of the others (hypothesis of niche conquest). The log-range and log-normal distribution are common for communities where species with average numbers become relatively more abundant and widely distributed (niches are multidimensional). The "broken shaft" model represents a situation, when species are distributed with maximally possible evenness (and hyper-capacity niche bounds are located randomly) [11, 12].

During revegetation of the northern slopes of the coal mining spoils open plant assemblage gradually had become more structurally complex: in the 2000s described by a geometric range, and in the following decades by a log-range and log-normal distribution. The model of "broken shaft" was found only once during the last year of observations on the spoils established in the 1990s (Fig.).

Therefore species composition of the plant assemblages developing during revegetation of the northern slopes of the coal mining spoils, i.e. in the habitats with the most favourable (in the arid zone)the slopes which in the arid zone conditions for plant growth and development, forms during the first 10-15 years of spoil revegetation.

All the species identified on the spoils belonged to three ecological groups: xerophytes, mesoxerophytes and mesophytes, with mesoxerophytes being most represented.

In contrast to the flora of stony and small tussock steppes of Khakassia, where endemic species represent *Fabaceae* family and *Astragalus* and *Oxytropis* genera, the latter were not found even after 40 years of coal mining spoils revegetation.

Gradual increase in the structure complexity of the open plant assemblages from the geometric range to the log-range and log-normal distribution was revealed. The "broken shaft" model was found only once during the last year of observations on the spoils dating back to the 1990s. In general all indicators evidenced that the structure of the studied plant assemblages was more or less developed/formed.

The study was financially supported by the Russian Foundation of Basic Research Grant 18-44-190006 r_a.

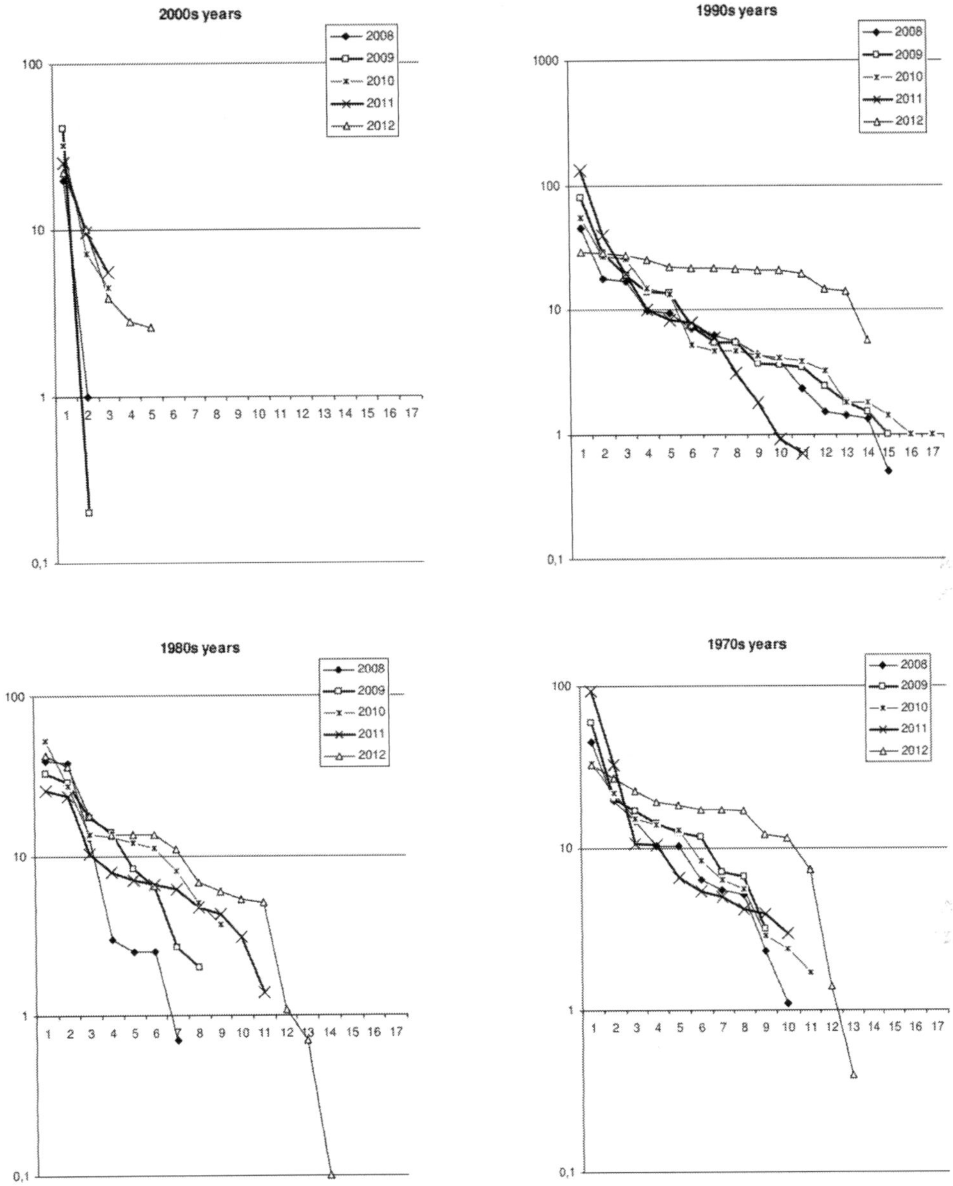

Figure. Significance curves for successive plant communities under natural revegetation of the northern slopes of the coal mining spoils in the arid areas of Khakassia.
Y axis represents species relative significances (vascular plants species air-dry phytomass, g/m^2).
X axis represents species ranks in descending order of abundances.

References

1. M. Hendrychová, J Landscape Studies **1**, 1 (2008)
2. R. Tropek, T. Kadlec, M. Hejda, P. Kocarek, J. Skuhrovec, I. Malenovsky, S. Vodka, L. Spitzer, P. Banar, M. Konvicka, Ecol. Eng. **43**, 1 (2012)
3. J. Banaszek, M. Leksy, O. Rahmonov, DOI: 10.3846/enviro.2017.098. (2017)

4. H. Lei, Zh. Peng, H. Yigang, Zh. Yang, Ecol. Eng. **94** (2016)
5. A.A. Korchagin, E.M. Lavrenko, *Field Geobotany* (USSR AS Pbs., Moscow-St.-Petersburg, 1976) in Russian
6. A.G. Voronov, *Geobotany* (Higher School, Moscow, 1973) in Russia
7. A.V. Kuminova, Y.M. Maskaev, G.A. Zvereva, Vegetation cover of Khakassia (Nauka Pbs., Novosibirsk, 1976) in Russian
8. T.G. Lamanova, in Geobotanical studies in West and Middle Siberia (Nauka Pbs., Novosibirsk, 1978) in Russian
9. A.V. Polozhy, Proc. SB USSR AS **4**, 1 (1964) in Russian
10. L.I. Malyshev, *The history of flora and vegetation in Eurasia* (Nauka Pbs., Novosibirsk, 1972) in Russian
11. R.H. Whittaker, *Communities and ecosistems* (MacMillan Pbs., NY., 1975)
12. E. Megarran, Ecological Diversity and Its Measurement. (Mir Pbs., Moscow, 1992)

BIO Web of Conferences **11**, 00037 (2018) https://doi.org/10.1051/bioconf/20181100037
Prospects of Development and Challenges of Modern Botany

Anatomic structure of leaf of some species of *Hemerocallis* genus

Lyudmila Sedelnikova [*], and *Lina* Cheltygmasheva

Central Siberian Botanical Garden, SB RAS, 630090 Novosibirsk, Russia

Abstract. This is research of anatomic features of structure of leaf of 4 species of *Hemerocallis* genus in the conditions of forest-steppe zone of Western Siberia. It is provided the description of upper and lower epidermis, cross section of leaf. It is revealed that the structure of lamina is characterized by xeromesophytic characters. It is mentioned species specificity in arrangement of adaxial and abaxial stomas pursuant thereto leaves of two types of daylily are amphistomatic and hypostomatic. The greatest number of stomas observed on the abaxial part of leaf (53 – 72 pieces per 1 sq.mm). The conductive system of leaf is presented by well developed bunches of collateral type alternating with large pneumatic cavities.

1 Introduction

Representatives of the genus *Hemerocallis* L. (daylily, trout lily, *Hemerocallidaceae* R. Br. family) are perennial polycarpic herbaceous rhizome plants. They are cultivated in botanic gardens worldwide from of old. It is known that culture conditions exert impact on changes of adaptive characters of plants. A leaf is the most plastic vegetative organ and in its structure reflected the ecological adaptation forming under the influence of species habitat conditions [1 – 3]. The aim is a comparative study of leaf anatomic structure of four species of *Hemerocallis* genus in the conditions of the forest-steppe zone of Western Siberia.

2 Material and methods

The research is performed in the Central Siberian botanic garden of the Siberian Branch of the Russian Academy of Science (Novosibirsk).As objects of research served leaves of 4 species: *H. citrina* Baroni, *H. fulva* L., *H. middendorfii* Trautv. et C. A. Mey., *H. minor* Mill. The leaves were collected from middle parts of bines at 2017 growing season. Cross sections were prepared by the sliding microtome with the freezing device. Specimens were studied and photographed by Axioskop-40 (Carl Zeiss) microscope with AxioCam MRc-5 color digital cameras of high resolution and the AxioVision 4.8 program. The description of lead anatomical structure was made by the standard techniques [4 – 6].

[*] Corresponding author: lusedelnikova@yandex.ru

3 Results and discussion

The leaves of the studied species were root, linear, sharp at their top, plain, with outstanding ribs, arcuate. The color of lamina varies from light green (*H. fulva, H. minor, H. middendorfii*) to dark green (*H. citrina*). Species differ on their leaves width, among increase in width daylilies were ranked as follows: *H. minor* (0.7 – 1 cm), *H. middendorfii* (1.8 – 2.2 cm), *H. citrina* (2.5 – 3 cm), *H. fulva* (3 – 3.5 cm).

The leaves have clear-cut upper (adaxial) and lower (abaxial) parts of lamina. Outside the leaf is covered with the epidermis characterized by compact arrangement of cells, existence of cuticle and stomas. There were noted each of general and individual traits in the structure of epidermal cells for all the species. The projection of the area of epidermal cells is rectangular (*H. citrina*) and quadrangular and hexagonal (*H. fulva, H. minor, H. middendorfii*), their outlines are rectilinear, their corners in allied borders are sharp and right (*H. citrina, H. fulva, H. minor*), more rarely they are roundish and blunt (*H. middendorfii*). There were revealed differences between cells of upper and lower epidermis. On the adaxial part of leaf the epidermis consists of large thin-walled cells, skintight to each other, they are extended, from rectangular to oval form (*H. fulva, H. minor, H. middendorfii*), *H. citrina* differs in more shortened cells with the curved borders. Earlier rectangular cells of the epidermis were noted in *H. fulva* [7]. The cell sizes are decrease, and cellular membranes are thickening in the direction from median vein to lateral face of lamina. The epidermis on the abaxial part of all the studied species consists of smaller compacted cells with reinforced walls. They are quadrangular or the hexagonal. Existence of larger epidermal cells on the adaxial part of leaf is an attribute of mesophyte plants. It is established that epidermal cells thickness of the adaxial part in the field of midrib (106.4 ± 11.2 – 162.2 ± 14.6 microns) is 3 – 4 times exceed on average thickness of cells of the abaxial part. At the same time, with distance from the center main rib, the thickness of upper epidermal cells is decreasing almost by 2 – 3 times. Thickness of abaxial cells decreases by 1.5 – 2 times on average. Except *H. middendorfii,* it was noted the identical sizes of adaxial epidermal cells. It was found the difference in the general thickness of leaf about its midrib. The greatest thickness is revealed for *H. fulva* (1245.5 ± 2.0 microns), the smallest one for *H. middendorfii* (851.4 ± 6.7 microns). The value of the general thickness of leaf about 3 – 4 veins vary in the range from 553.5 ± 22.2 (*H. middendorfii*) to 734.5 ± 26.4 microns (*H. minor*). Stomas are oval and they are located at epidermal cells junctions. They are distributed by the ranks parallel to longitudinal axis of leaf and they are dip into epidermis. The stomatal apparatus of tetracytic type, guard cells are surrounded with four accessory cells - two of them are lateral ones and other two are polar. Such type is characteristic of monocotyledonous plants. However, some researchers [3, 8] noted the presence of anomocyte type of stomata in daylilies. For *H. citrina* and *H. minor* we were found abaxial and adaxial stomas (amphistomatic leaf). There are more abaxial stomas (tab. 1) that allows reducing leakage of water during transpiration and enhancing gas exchange.

Table 1. Number of stomata per 1 mm^2 of leaf surface

Species	stomata per 1 mm^2 of leaf surface, pcs. M ± m	
	Upper epidermis	Lower epidermis
H. citrina	42.8 ± 3.1	59.4 ± 4.5
H. fulva	-	53.0 ± 4.3
H. middendorfii	-	72.5 ± 5.2
H. minor	31.6 ± 3.3	54.2 ± 10.3

Note. M is the arithmetic mean, m is the standard error.

In upper epidermis their number varies from 31 to 42, and on the lower from 54 to 59 per 1 mm^2. Leaves *of H. fulva* and *H. middendorfii* are characterized by existence of stomas only on the lower epidermis (hypostomatic leaf). Their number changes in the range from 53 to 72 per 1 mm^2, with the greatest dominance for *H. middendorfii.* Submersed stomas and their existence on the both sides of leaf are connected with xerothytic habitat conditions.

All space between the lower and upper epidermis is filled with the mesophyll containing large amount of chloroplasts. It is presented by the spongy cell tissue consisting of rather small, horizontally extended oval cells, adherent to each other. Cells of palisade parenchyma from upper epidermis are single-layered and expressed poorly, in comparison with the data of [3], where the daylilies grew in arid conditions. On the adaxial part of the studied daylilies about midrib there are from 2 (*H. fulva, H. minor)* to 3 rows of parenchyma cells (*H. middendorfii, H. citrina*). On the abaxial part of the majority of species *(H. minor, H. middendorfii, H. citrina)* there were noted 4 – 5 rows, for *H. fulva* there is 5 – 6 rows. With distance from the main vein on the lower part of leaf there was observed reduction of parenchyma cells layers down to 2 – 3 (*H. minor, H. middendorfii, H. citrina)*, except *H. fulva* (4 layers) (see fig.1).

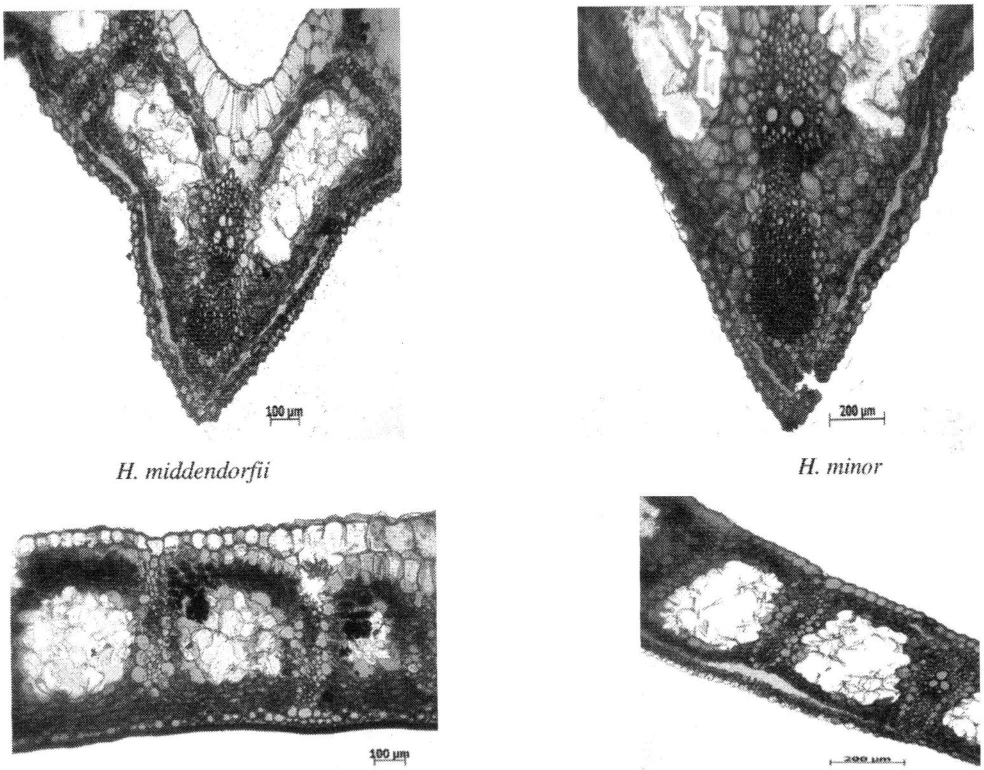

Fig. 1. Anatomical structure of transverse section of *Hemerocallis* leaf

The leaf mesophyll is penetrated densely by the veins forming uniform conductive system. For all the species there are well-marked central and the side conducting bundles having ordered arrangement in the same row, in parallel each other, at the same time large bunches belong to midrib, and with distance from it to both parts of leaf blade the sizes of conducting bundles are decreasing. In bunches the xylem is oriented to the adaxial party of leaf, and phloem – to abaxial. At such orientation leaf conducting tissues are naturally closed with stalk conducting tissues. Xylem vessels the large and well developed and phloem cribriform cells are smaller. Conducting bundles are closed collateral on their type. From the upper and lower parts of conducting bundles are located sclerenchyma parietal cells, which have quadrangular and hexagonal shape and thick membrane. The sclerenchyma protects bunches from mechanical damages and gives durability to leaf blade in general. The conducting bundles alternate with large pneumatic cavities. For different species pneumatic cavities have the different shapes: trapezoid (*H. minor*) and rectangular (*H. fulva, H. middendorfii, H. citrina*); that was not noted for strains earlier [9]. The arrangement of conducting bundles among mesophyll and pneumatic cavities promotes optimum performance of photosynthetic and transport system. Analysis of our data with literature data [3, 7 – 9] showed how common features in leaf anatomic structure of *Hemerocallidaceae* family, and the presence of species features – tetracite type of stomata, mesophytic and xerophytic features. Adaptations in *H. fulva, H. minor, H. citrina* have been revealed, expressed in the quantitative change and structure of mesophyll cells, air cavities, conducting beams in these species under different conditions, which indicates ecological plasticity and the ability to adapt them to wet and dry habitats.

4 Conclusion

The studied daylilies possess amphistomatic(*H. citrina, H. minor*) and hypostomatic (*H. fulva, H. middendorfii*) leaves of dorsoventral type with the well-marked adaxial and abaxial parts. The greatest number of stomas is revealed on the lower epiderma of leaf (53 – 72 per 1 mm^2). Conducting bundles are closed collateral; they alternate with large pneumatic cavities. Leaf blades combine mesoxerothytic characters that is connected with their origin, promotes plasticity and well-marked adaptation of the studied species to conditions of forest-steppe zone of Western Siberia.

This work is performed within the state task of the Central Siberian Botanic Garden of the Siberian Branch of the Russian Academy of Science on the project № AAAA-A17-1170126100053-9 "Identification of ways of plants adaptation to contrast habitat conditions at the population and organismal levels". During the preparation for the publishing there were used materials of the CSBS Siberian Branch of the Russian Academy of Science bioresource scientific collection "Collections of live plants on the field and glass-covered ground", UNU № USU 440534.

The authors express their gratitude to A. A. Krasnikov for the methodical help in learning the equipment.

References

1. Yu.V. Gamaley, *Leaf phloem* (Nauka, Leningrad, 1990)
2. LF. Hen, PD., Jiang, Y. Wang, HZ. Liu, Northern Horticulture **23**, 71 (2011)
3. I.I. Krokhmal, Ecology and Noospherology **23**, 41 (2012)
4. E.A. Miroslavov, *Structure and function of the epidermis of the leaf of angiosperms* (Nauka, Leningrad, 1974)
5. K. Ezau, *Anatomy of seed plants* (Mir, Moscow, V.2, 1980)
6. CY. Zhou, M. Nie, CD. Yang, Journal of Yangtze University (Natural Science Edition) **3,** 44 (2012)
7. W.B. Zomlefer, Harvard Papers in Botany **3,** 123 (1998)
8. L.L. Sedelnikova, Vestnik KGAU **7**, 55 (2016)

Biogeocenosis development during initial revegetation of a coal combustion ash dump

Natalia Sheremet[1,*], *Ivan* Belanov[2], *Vladimir* Doronkin[1], *Tatiana* Lamanova[1], and *Natalia* Naumova[2]

[1]Central Siberian Botanical Garden, SB RAS, 630090 Novosibirsk. Russia
[2]Institute of Soil Science and Agrochemistry, SB RAS 630090, Novosibirsk, Russia

Abstract. The initial stage of biogeocenoses development on the coal ash dump produced by the thermal power staton in Novosibirsk (55.000, 83.068), Russia, were studied after 9 years of spontaneous revegetation. Soil properties, soil cover and plant communities were examined in detail. The predominating types of embryozems and transition from open to succession plant communities were described. Soil substrate moisture content was found to determine changes in plant species composition, projective cover and abundance, altogether causing asynchronicity of soil formation in different sites.

1 Introduction

Several major power plants operate in Novosibirsk, combusting coal to produce electric power and heat for the city. Of these plants Central thermal power station-5 (CTPS-5) is the biggest in the city and one of the biggest in the Asian part of Russia, accounting for 50% of electric power production in the Novosibirsk region.

According to the technological process, the coal combustion wastes (fly ash, an oxidised, non-combustible material of very fine size and grayish in colour) are fluidized and transported into ash dumps. After the dumps are filled to their projected capacity, they require certain recultivation for the occupied territories to be safely used. Such technogenic dumps can be sources of adverse effect [1, 2] on the ground water and air quality due to diffusion of chemical compounds and ash spreading by wind erosion, respectively. Thus assessment of ecological status and determination of the rate of soil and vegetation natural restoration are quite urgent for understanding the ways to decrease adverse environmental impacts [3].

Ash dump 1 from the CTPS-5 is located upwards in Baryshevsky ravine near the watershed between the Inya and Plyushikha Rivers (55.000, 83.068). The dump was started in 1987, filled to the projected capacity in 2008 when the transportation of the fluidized ash was stopped. The dump of the total area of 41.1 ha consists of two areas divided by a dam. In 2010-2011 one area was subjected to technical recultivation by overlaying the ash with a layer of potentially fertile substrate of varying thickness. The other part of 17.7 ha was not overlaid and left for natural revegetation.

* Corresponding author: nsheremet@yandex.ru

The aim of the work was to characterize in detail soils and phytocenoses development during the initial stages of revegetation of the coal ash dump of the CTPS-5 in Novosibirsk.

2 Materials and methods

2.1 Location of the study sites

The study area is located 55.000, 83.068. The study was carried out on three coal ash dump sites during the 9th year under natural revegetation. The sites were adjacent to each other. The area of the studied sites was 4 ha for Site-1, 6.2 ha for Site-2 and 5 ha for Site-3. According to the regional agroclimatic zoning, the Novosibirsk region falls within the forest-steppe zone of the Northern Cis-Altai province. The climate of the region is characterized with the sum of above +10°C air temperatures of 1926°C·day, yearly precipitation of 300-400 mm, 119 days of frost-free period and the average daytime temperature of July (the hottest month) of 25.7 °C.

The studied revegetating coal ash dump sites are adjacent to the small areas covered by birch (*Betula pendula* Roth.) forest with goutweed and grasses as surface cover and the underbrush of *Sorbus sibirica* Hedl., *Viburnum opulus* L. *Crataegus sanguinea* Pall. and *Pinus sylvestris* L.

Since the study sites had not been recultivated, the plants disseminated on the pure mineral substrate, i.e. coal ash, thus the revegetation successions can be considered as primary. Coal ash is a specific mineral substrate, impeding plant growth due to high pH and the absence of nitrogen.

Soil and plant sampled were collected in July and August 2017.

Soils were described according to the soil substrate genesis classification of technogenic landscapes [4] as embryozems or, according to WRB, as technosols [5]. Zonal soil types in the forest-steppe of the region are grey wooded soils, or Haplic Luvisols [5].

According to the Russian classification, the main technosol types at the initial stages of soil formation are called embryozems, which can be grouped in the following main types: initial embryozem (Ei), organic matter accumulating embryozem (Eo), soddy embryozem (Es) and humus accumulating embryozem (Eh). Embryozem development goes from Ei—Eo–Es–Eh syngenetically with plant succession stages [6].

At each study site a pit was dug, and 3 subcores were taken from nascent genetic horizons, bulked together to comprise one composite sample. Field-moist soil samples were 2-mm sieved and stored in a refrigerator (+4 °C) prior to analyses. The content of soil organic (SOC) and soil inorganic carbon (SIC) was determined by stepwise loss on ignition method [7] using 2-4 g soil aliquots. Soil total nitrogen (STN) was determined by Perkin-Elmer elemental analyzer. Available forms of macronutients (NO_3^-, NH_4^+, P_2O_5) were determined in field-moist samples by standard techniques: briefly, nitrate was determined potentiometrically in 0.03M K_2SO_4 extracts, while ammonium was measured colorimetrically in 2N KCl extracts, and available P was extracted with 0.5 M $NaHCO_3$ solution and determined colorimetrically [8]. Soil pH was measured in a supernatant of soil-water solution (1:5 v/v). Exchangeable K^+, Na^+, Ca^{2+} and Mg^{2+} were determined by atomic adsorption in ammonium citrate extracts. Granulometric composition of soil substrate was determined according to [9]. Total dissolved solids in soil water extracts were measured potentiometrically. The rate of carbon dioxide release by soil substrates, i.e. soil respiration, was measured in laboratory with CO_2 data loggers.

To study the floristic composition of plant communities plant samples for herbarium were collected and geobotanical descriptions made in each key study site during the growing season. To determine the above-ground phytomass standing stock and species contribution into it the aboveground plant material was clipped by species in 4 replicates, air-dried and weighted.

Soil properties were analyzed by principal component analysis with *Statistica v.6.1* software.

3 Results

3.1 Site-1

Revegetation and soil development on Site 1 resulted in 80% of the area occupied by Ei and 20% occupied by Eo. Since Ei in their physical and chemical properties are very similar to the original ash substrate, substantial contribution of Ei areas into soil cover can be assessed as edaphically not favourable environment for phytocenosis development. Granulometrically both Ei and Eo soils were light loams with alkaline pH (8.02–8.83). The 0-20 cm soil layer had low field water content (8.3–10.5%) when sampled. The C_1 horizon had soil dry mass density of 1.3 g/cm^3, while C_2 horizon had 0.9 g/cm^3. Total porosity among structural aggregates accounted for 49.2–63.8%. Mass density of the solid matter on this site was the highest, reaching 2.6 g/cm^3, whereas organic matter content in the root-penetrated layer was the lowest among the studied embryozems (*ca.* 1.5%).

The 9 years of Site-1 natural revegetation resulted in the open herbaceous and smallreed assemblage, accompanied by *Hippophaë rhamnoides* L. and with *Calamagrostis epigeios* (L.) Roth. predominating. The total projective cover was 5%. Higher vascular plants were represented by 18 species. The vegetation consisted of singular plants, growing on the smaller part of the Site-1 area. The grass stand was on average 40 cm high. The herb assemblage was dominated by sages *Artemisia integrifolia* L. and *A. vulgaris* L., tower rockcress (*Turritis glabra* L.). The legumes were dominated by white sweet clover (*Melilotus albus* Medic.). Woody plants were represented by the buckthorn *Hippophaë rhamnoides* L. pyramid poplar *Populus italica* (Du Roi) Moench underbrush. Bryophytes also made important contribution into the site plant assemblage with their projective cover of 70–80% and 4 species, among which *Leptobryum pyriforme* (Hedw.) Wits. and *Bryum sp.* predominated.

The standing aboveground living phytomass was estimated to reach 109 kg/ha, while standing dead phytomass was 9 kg/ha.

3.2 Site-2

On this site Ei soil first transformed into Eo and then further into Es due to the plant community development from the open to the closed one, dominated by grasses. Revegetation and soil development on this site resulted in 75% of the area occupied by Es and 25% occupied by fragmented Eo. Soil environment is quite favourable for plant growth and development. Granulometrically both Es and Eo soils were medium loams with alkaline pH (8.08–8.65). Soil substrate moisture content is an important factor, determining plant community structure on revegetation coal ash dumps. As compared with Site-1, Site-2 soil substrates contained more water (27–32%). The A_1–C_1 horizons were more loose and less dense (0.4–0.7 g/cm^3), with C_2 horizon being slightly more dense (0.9 g/cm^3). Total porosity among structural aggregates changed from 80 to 60% with soil depth. Mass density of solids on this site was 2.2 g/cm^3, whereas organic matter content in the root-penetrated layer was the highest among the studied embryozems, ranging 2.2–3.7%.

The clover and smallreed phytocenosis had formed on Site-2 over the 9 years of natural revegetation. The aboveground phytomass stand was rather dense with an average height of

90 cm and 90–100% of the total projective cover. The main gramineous and leguminous edificator plants were represented by *Calamagrostis epigeios* and *Melilotus albus*, respectively. Relatively even distribution of *C. epigeios* throughout Site-2 area resulted from its biology, i.e. was due to the species ability to disseminate rapidly via rhizomes. High abundance of white sweet clover plants, observed on Site-2, was observed earlier on Verhny Taguil coal ash dump in the Urals region [10]. In contrast to evenly distributed smallreed plants, the clover plants had patchy distribution, resulting in mosaic vegetation. The plant community was found to consist of 14 higher vascular plant species and 3 pryophyte (*Barbula unguiculata* Hedw., *Bryum cf. caespiticium* Hedw., *Ditrichum pusillum* (Hedw.) Hampe).

The standing aboveground living phytomass was estimated as 4961 kg/ha, while standing dead phytomass was 1154 kg/ha.

3.3 Site-3

The soil cover of the site was equally comprised by Eo and Es. Granulometrically both Eo and Es embryozems were medium loams with alkaline pH (8.23–8.93). The 0-20 cm soil layer had low field water content (8.3–10.5%) when sampled. The C_1 horizon had specific mass of 1.3 g/cm^3, while C_2 horizon had 0.9 g/cm^3. As compared with Site-1 and Site-2, Site-3 soil substrates contained excessive water (more than 39% in 0-20 cm layer). Similarly to Site-2, the A_1–C_1 horizons were more loose and less dense (0.4–0.7 g/cm^3), with C_2 horizon being slightly more dense (0.9 g/cm^3). Total porosity among structural aggregates changed from 60 to 45% with soil depth. Mass density of solids in Site-3 soil substrate was 2.1 g/cm^3, whereas organic matter content in the root-penetrated layer was the highest among the studied embryozems, ranging 1.2–2.4%.

Due to higher soil substrate water content on Site-3 there grew such hydrophytes as the graceful cattail *Typha laxmannii* Lepech. and the common reed *Phragmites australis* (Cav.) Trin. ex Steud. The Site-3 area is intermittently flooded. The phytomass stand was on average 80 cm high, forming 40-50% of total projective coverage. The phytocenosis could be described as the smallreed-bentgrass one. Similar to Site-2, the main edificator was *Calamagrostis epigeios*, with its bunch stabilizing ash substrate and hence decreasing dust spreading by wind. The dominating redtop bentgrass *Agrostis gigantea* Roth. was spread unevenly throughout the Site-3 area. Grasses dominated the site plant community composition. Woody plants were represented by two willow species (*Salix caprea* L., *S. viminalis* L.), their trees reaching as high as 2 m sometimes.

The standing aboveground living phytomass was estimated as 2373 kg/ha, while standing dead phytomass was 1673 kg/ha.

3.4 Principal component analysis

Principal component analysis of the data matrix with soil properties as variables and different study sites as objects showed quite distinct location of the studied sites in the plane of PC1 and PC2 (Fig.1), together accounting for more than 2/3 of the soil data variance.

The humus horizon of the undisturbed Luvisol under birch forest, as expected, lies far apart from the technosols due to organic C, total and total contents, as shown by the location of soil properties in the plane of PC1 and PC2 (Fig.2) As PC1 can be interpreted as related to soil substrate physical properties mostly, while PC2 is related to organic matter quantity and quality. Soil respiration rate also contributed to PC1, being higher in Site-2 and Site-3 soil substrates due to increased fresh organic carbon input with increased phytomass production and higher water content.

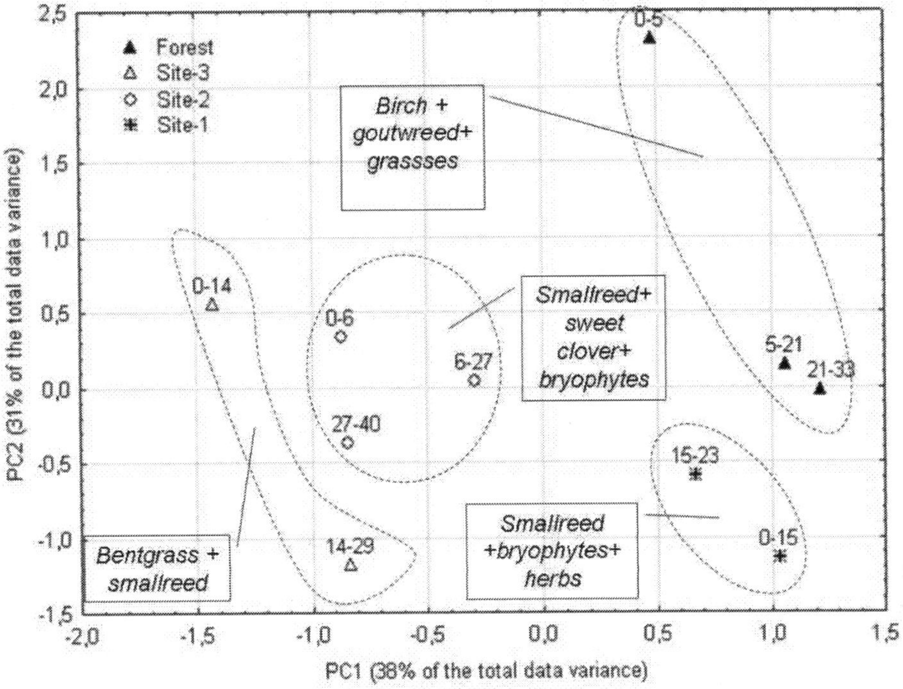

Fig.1. Location of soil samples in the plane of the first two principal components. Forest – undisturbed birch forest on Haplic Luvisol; Site-1, Site-2 and Site-3 – sites on the coal ash dump under natural revegetation for 9 years on Technosols. Numbers above markers indicate soil substrate sampling depth in cm. Text in rectangles describes plant dominants.

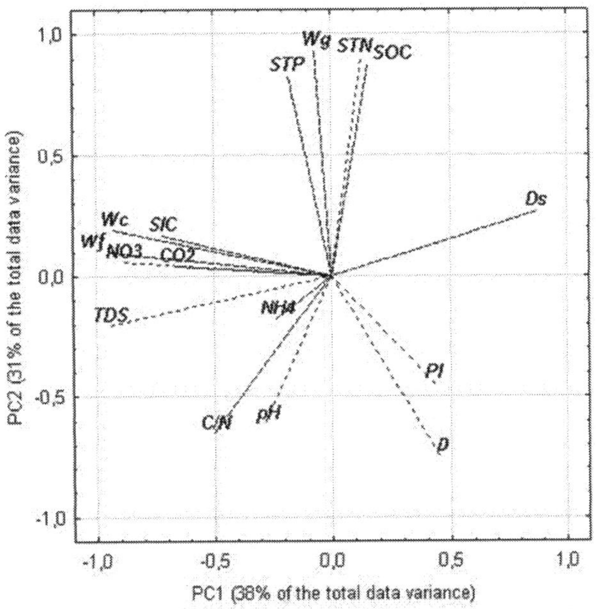

Fig.2. Location of soil substrate properties in the plane of the first two principal components. Abbreviations: STP – soil total phosphorus, STN – soil total nitrogen, SOC – soil organic carbon, CO_2 – soil respiration, SIC – soil inorganic carbon, TDS – total dissolved solids, Wc, Wg, Wf – capillary, hygroscopic and field water contents, resp., D-soil density, Ds – density of solids, Pl – labile phosphorus content.

4 Conclusion

The results showed that on all studied sites of the coal ash dump 9 years of natural revegetation of the ash substrate led to the phytocenoses, dominated by perennial herbs and grasses (25 species) and scarce woody plants (4 species). Bryophytes were represented by 7 species. Increasing water in the topsoil substrate resulted in shifts in vegetation and soil cover from the open plant community on the initial embryozem to grasses and legumes community on the organic matter accumulating embryozems and further on towards the grasses community on the organic matter accumulating and soddy embryozems. On the two study sites environment favoured phytomass production and biological turnover in general and, consequently, soil development, resulting after nine years in less than 5% of soil cover occupied by initial embryozem. Changes in water content of the coal ash substrate brought shifts in plant species composition, projective cover, plant abundance and phytomass production, altogether leading to asynchronicity of site-specific soil formation stages. Thus despite the apparent monotony of the coal ash dump (horizontally even surface, homogenous substrate with the same properties at the start of revegetation) different ecotopes developed due to specific environmental conditions, such as microrelief, local water input and retention. The ecotope diversity determined vegetation diversity and phytomass production and hence soil formation differentiation. Overall it can be concluded that natural revegetation of coal ash dumps leads to productive plant communities under sufficient water supply.

The study was financially supported by Grant from the Russian Fund of Fundamental Research (№ 18-44-540002 p_a). Bryophyte species were determined by Doctor of Biological Sciences O. Yu. Pisarenko from the Central Siberian Botanical Garden of the SB RAS, 630090, Novosibirsk, Russia

References

1. R.J. Haynes, J. Environ. Manage. **90**:1, 43 (2009)
2. S.Y. Pan, H. Morrison, L.Gibbons, J.Zhou, S.W.Wen, M.Desmeules, Y.Mao, J. Occup. Environ. Med. **53**:5, 522 (2011)
3. D.Maiti, B. Prasad, Appl. Ecol. Env. Res. **14**:2, 185-212 (2016)
4. V.M. Kurachev, V.A. Androkhanov. Contemporary Problems of Ecology. **3** (2002). In Russian
5. IUSS Working Group, *WRB, World Reference Base for Soil Resources 2006. International soil classification system for naming soils and creating legends for soil maps* (FAO, Rome, 2006)
6. V.A.Androkhanov, V.M.Kurachev, E.D.Kulyapina, *Soils of technogenic landscapes: genesis and evolution,* (SB RAS Pubs., Novosibirsk, 2004) in Russian
7. Q. Wang, Y.Li, Y. Wang, Environ. Monit. Assess. **174,** 241-257 (2011)
8. W.H.Hendershot, H. Lalande, M.Duquette, *Soil Sampling and methods of Analysis* (CRC Press, Boca Raton, 2008)
9. D.Kroetsch, C.Wang, *Soil Sampling and methods of Analysis*. (CRC Press, Boca Raton, 2008)
10. A.K. Makhnev, T.S. Chibrik, M.R. Trubina, N.V. Lukina, N.E. Gebel, A.A. Terin, Yu.I. Elovikov, N.V. Toporkov, *Ecological basis and methods for biological recultivation of coal ash from power heating stations in the Ural region* (Ekaterinburg, 2002) in Russian

BIO Web of Conferences **11**, 00039 (2018) https://doi.org/10.1051/bioconf/20181100039
Prospects of Development and Challenges of Modern Botany

Ontogenetic structure of *Thymus* L. (*Lamiaceae*) coenopopulations in Yakutia

Evgeniya Talovskaya*

Central Siberian Botanical Garden SB RAS, Zolotodolinskaya st., 101, Novosibirsk, Russia, 630090

Abstract. Ontogenetic structure of coenopopulations of four model species *Thymus extremus*, *T. indigirkensis*, *T. brevipetiolatus* and *T. pavlovii* widely distributed in Yakutia was studied. It was established that species coenopopulations were normal, most of them incomplete. Dependence of the type of the ontogenetic spectrum of coenopopulations on growth conditions and ontogenesis pattern was revealed in thymes of different biomorphs. The bimodal type of the ontogenetic spectrum (monocentric dwarf subshrub *T. extremus*) is conditioned by irregular seed regeneration and lack of ripe and old generative stages. Formation of the left-side type of the ontogenetic spectrum is governed by the emergence of vegetative propagation of individuals (unobvious polycentric dwarf subshrub T. *brevipetiolatus* and dwarf semishrub *T. indigirkensis*) and substrate features (distinct polycentric dwarf semishrub *T. pavlovii*).

1 Introduction

Currently searching for ways of rational nature management has become increasingly important [1, 2]. A number of issues along this line are devoted to study of natural reserves of medicinal plant material. *Thymus* L. species are widely used in both official and common medicine. Mass collection of thymes by local population leads to sharp deterioration of natural population status and, as a consequence, to decrease in natural resources. One of the criteria of assessment of the current status of coenopopulations of medicinal plant species is their ontogenetic structure [3]. Medicinal properties of *Thymus* species are known since ancient times. They possess disinfecting, analgesic and antiseptic activities. *T. extremus* Klok., *T. indigirkensis* Karav., *T. brevipetiolatus* Čáp and *T. pavlovii* Serg. are widespread in the territory of Yakutia. The main habitats of the species are steppe plant communities on the southern and south-western cobble slopes of river banks and tops of the hills [4]. In literature there are data on the biology of *Thymus* sibiricus (Serg.) Klokov et Shost. growing in Central Yakutia [5], whereas data on the ontogenetic structure of coenopopulations of this and other species are absent. In this regard, the aim of the work was study of the ontogenetic structure of coenopopulations of model species *T. extremus*, *T. indigirkensis*, *T. brevipetiolatus*, and *T. pavlovii* in Yakutia.

* Corresponding author: kolegova_e@mail.ru

2 Materials and Methods

Coenopopulations (CP) of four model species *T. extremus*, *T. indigirkensis*, *T. brevipetiolatus*, and *T. pavlovii* growing in different ecological-coenotic conditions of Yakutia were studied (Table 1). When studying the ontogenetic structure of CP, standard methods were used [6–8]. The transects 1 m wide and 5–10 m long were divided into test plots at 1m^2, a continuous account of individuals at each of the ontogenetic stage was carried out. As a unit of account, a primary shoot, a primary bush and a clump in individuals of seed origin, a partial bush and a system of partial bushes in individuals of vegetative origin were taken into consideration. An ontogenetic spectrum of CP was defined as a ratio of plants at various ontogenetic stages expressed in a percentage of the total individual number [6]. To characterize CP were used demographic indices of age and effectiveness. The type of a coenopopulation was established by L.A. Zhivotovskiy's "delta-omega" classification [7]. The ecological density of CP was defined when calculating individual abundance per unit of habitable space [8].

Table 1. Characteristic of habitats of *Thymus* coenopopulations in Yakutia

CP	Locality of CP	Community / dominating species	TPCD/PCD %
1 (*T. extremus*)	Vicinities of Yakutsk, gentle slope of the Lena River bank, south-eastern exposure, steepness 10^0, meadow-chernozem sandy-loam soil	Herb-sheep's fescue steppe (*Festuca lenensis* Drobow, *Artemisia pubescens* Ledeb., *Veronica incana* L., *Carex duriuscula* C.A. Mey., *Pulsatilla flavescens* (Zucc.) Juz.)	80/3
2 (*T. brevipetiolatus*)	Kobyansk region, vicinities of s. Smorodichny, slope of southern exposure, steepness 5^0, loam with cobble.	Spreading anemone-sheep's fescue-wildrye steppe (*Festuca lenensis, Elymus turuchanensis* (Reverd.) Czerep., *Pulsatilla patens* (L.) Mill., *Koeleria cristata* (L.) Pers., *Veronica incana*)	80/1
3 (*T. indigirkensis*)	Kobyansk region, vicinities of s. Sangar, slope of southern exposure of the Lena River bank, steepness 5^0, fine grain soil covered with cobble	Bush encroached by *Spiraea dahurica* (Rupr.) Maxim. and *Rosa acicularis* Lindl. herb bunchgrass steppe (*Agropyron villosum* (L.) Link, *Festuca lenensis, Carex pediformis* C.A. Mey., *Dracocephalum palmatum* Steph. ex Willd., *T. indigirkensis*)	60/5
4 (*T. pavlovii*)	Tompo region, southern exposure of	Grouping of vegetation (*Bromopsis pumpelliana*	35/5

	the slope of the Sete-Dabash Ridge, right борт of the Tyi-Synoga stream valley, steepness 30^0, rocky-subsiding outcrops, moss in crevices	(Scribn.) Holub, *Elymus confusus* (Roshev.) Tzvelev, *T. pavlovii*, *Dracocephalum stellerianum* F. Hiltebr., *Rhytidium rugosum* (Hedw.) Kindb.)	

Note: TPCD – total projective cover degree of the herb layer, PCD – projective cover degree of the species.

3 Results and Discussions

Study of *T. extremus*, *T. indigirkensis*, *T. brevipetiolatus* and *T. pavlovii* showed that in Yakutia peculiar morphological mechanisms of adaptation manifesting themselves in the diversity of life forms and biomorph types were formed in the species. In CP 1 *T. extremus* is a monocentric dwarf semishrub. The phase of a primary bush is prolonged in the individual ontomorphogenesis. Self-maintenance of CP occurs only by seeds. In CP 2 *T. brevipetiolatus* and in CP 3 *T. indigirkensis* are implicitly polycentric dwarf semishrub and dwarf semishrub, respectively. The most prolonged phase in the individual ontomorphogenesis is that of a clump. Self-maintenance of CP occurs by seeds and vegetatively. IN CP 4 *T. pavlovii* is explicitly dwarf semishrub. An adult individual also represents a clump. Self-maintenance of CP occurs vegetatively. Studied CPs of model species are normal, mostly incomplete (subsenile individuals are absent) (Table 2).

Table 2. Distribution of *Thymus* species individuals throughout the ontogenetic groups

CP	Ontogenetic groups, %								$D_{ecol.}$	ω	Δ	CP type
	j	im	V	g1	g2	g3	ss	s				
1	12.2	17.1	7.3	17.7	15.2	9.8	20.1	0.6	23.4	0.55	0.42	transitional
2	0.4	6.3	15.7	33.2	16.2	19.7	8.5	0	40.0	0.70	0.41	transitional
3	3.8	5.0	20.0	38.1	13.2	15.0	4.9	0	74.3	0.67	0.35	transitional
4	13.2	36.8	17.6	19.1	5.9	1.5	5.9	0	8.5	0.39	0.19	Young

Note. Ontogenetic groups: j – juvenile; im – immature; v – virginal; g1 – young generative; g2 – ripe generative; g3 – old generative; ss –subsenile; s – senile; D_{ecol} – ecological density (individual/m^2); ω –effectiveness index; Δ - age index; type of coenopopulations according to L.A. Zhivotovskiy's classification (2001).

The ontogentic spectrum of CP 1 *T. extremus* is bimodal. An absolute maximum is in the right-side part of the spectrum on subsenile individuals. Two local maxima are distinguished on immature and young generative individuals in the left-side part of the spectrum. Multiple top of the spectrum is conditioned by different factors. On the one hand, irregular seed regeneration encourages accumulation of different groups of regrowth. On the other hand, accumulation of subsenile individuals is linked with features of *T. extremus* development in forb-sheep's fescue steppe community with a lack of water supply and high sodding. So young individuals pass to the subsenile stage omitting ripe and old generative ones. In this regard, the group of ripe generative individuals in CP 1 does not prevail. According to "delta-omega" classification the CP is transitional.

The ontogenetic spectra of CP 2–4 are left-side, with the maximum on young generative (CPs 2 and 3) and immature individuals (CP 4). Formation of the left-side ontogenetic spectrum in CPs 2 and 3 in *T. brevipetiolatus* and *T. indigirkensis* is similar. Both CPs occur in the steppe communities on hill slopes. Vegetative spreading and propagation are typical of species individuals in these conditions. Vegetative propagation occurs at the ripe generative stage. Ramets rejuvenated up to the young generative stage and sometimes in *T. indigirkensis* up to the virginal one annually separate from the maternal individual. This, as well as seed regeneration promote increase in young generative individual group and formation of the peak in the left part of the spectrum. Ecological density of CP of implicitly polycentric thyme species is high and reaches in *T. indigirkensis* 74,3 individual/m^2. CP self-maintenance is equally by seed and vegetative ways. Genet-ramet ratio is practically the same: in *T. brevipetiolatus* CP it amounts to 1:1,4, in *T. indigirkensis* CP – 1:1,2. According to "delta-omega" classification CPs 2 and 3 are transitional to ripe ones.

In *T. pavlovii* a rise in the left part of the spectrum falls on the group of immature individuals. The species habitat is characterized by specific features of the substrate. Individuals develop on the moss cover in the crevices of rocky outcrops. Favourable conditions for seed germination are formed here. However, only a small quantity of individuals passes to virginal stage in the future. This is linked with an adverse effect of moss in individual life at the later ontogenetic stages (conservative of permafrost, assists in preservation of soil low temperatures, leads to swamping). The same thing is also noted by J. L. Gornal et al. [9]. Duration of the ontogenesis of *T. pavlovii* genets does not exceed 8 years. At the ripe generative stage primary individual structures decay, and a plant breaks up completely into even-aged partial formations. CP self-maintenance is mainly due to vegetative regeneration of ripe generative individuals. Genet-ramet ratio in CP is 1:3,5. According to "delta-omega" classification CP 4 is young.

Thus, the study of the coenopopulations of model *Thymus* species in various ecological-coenotic conditions of Yakutia shows that all of them are normal, most of them are incomplete (individuals at the senile ontogenetic stages are absent). Dependence of the ontogenetic spectrum type on growth conditions and ontogenesis pattern was revealed in species of various biomorphs. Similar dependence was also revealed by some researchers when studying plant population structure in certain regions [3, 10 and others.]. A bimodal type of the ontogenetic spectrum in the monocentric dwarf semishrub *T. extremus* is governed by irregular seed regeneration and omitting ripe and old generative stages. Formation of the left-side ontogenetic spectrum of CP in implicitly polycentric semishrub *T. brevipetiolatus* and dwarf semishrub *T. indigirkensis* is conditioned by emergence of individual vegetative propagation, ramet rejuvenation (up to the young generative stage), combined way of CP self-maintenance. Accumulation of regrowth and formation of the left-side spectrum in explicitly polycentric dwarf semishrub *T. pavlovii* are to a greater extent due to peculiarities of moss substrate.

The work was carried out with the financial support of the grant of RFBR within the framework of project № 18-04-00621 and State Assignment of Central Siberian Botanical Garden SB RAS № AAAA-A17-117012610053-9

References

1. O.V. Smirnova, N.A. Toroova, Advances in current biology **136**, 199–211 (2016)
2. S.-L. Chen, H. Yu, H.-M. Luo, Q. Wu, C.-F. Li, A. Steinmetz, Chin. Med. **11**, 37–44 (2016)
3. O.V. Simonova, *Principles and methods of biodiversity conservation* (MarSU, Yoshkar-Ola, 2010) (In Russian)
4. V.M. Vasjukov, Bot. J. **101,** 1240–1253 (2016)
5. P.V. Egorova, Bul. of KrasGAU **7,** 65–69 (2014)

6. O.V. Smirnova, L. B. Zaugol'nova and etc, *Cenopopulation of plants.* 217 (1976) (In Russian)
7. L.A. Zhivotovsky, Rus. J. of Ecol. **1**, 3–7 (2001).
8. Y. Odum, *Ecology* (Moskow, 1986)
9. J.L. Gornal, S.J. Woodin, I. S. Lonsdottir, R. Van der Wal, Oecologia **166**, 769–782 (2011)
10. L.C. Bernacci, Acta Botanica Brasilica **22**, 119–130 (2008)

BIO Web of Conferences **11**, 00040 (2018) https://doi.org/10.1051/bioconf/20181100040
Prospects of Development and Challenges of Modern Botany

Ecological and Phytosanitary Monitoring of Indoor Air Quality in Preschool Institutions of Novosibirsk

Natalia Tsybulya[1,*], *Elvina* Fedorenko[2], and *Liliya* Gribkova[3]

[1]Central Siberian Botanical Garden, Novosibirsk, Russia
[2]Tomsk State University, Tomsk, Russia
[3]Kindergarten No. 195, Novosibirsk, Russia

Abstract. Special assortment of plants demonstrating high antimicrobial activity was tested in the play rooms of kindergartens in Novosibirsk. Preliminary laboratory experiments revealed universal activity of these plant species against test objects: *Staphylococcus epidermidis, Escherichia coli, Candida albicans.* Microbial pollution of air in the play rooms of 12 kindergartens in Novosibirsk was monitored taking into account the presence or absence of plants in the rooms. The dynamics of microbial pollution of the rooms was studied: the composition and number of colony-forming units per 1 m^3 (CFU/m^3), the fractions of opportunistic pathogenic microfrlora (staphylococcus, enterobacteria). The plants with the leaf area 1.5-3.0 m^2 per 100 m^3 of the room volume caused a significant effect of volatile exo-metabolites: a decrease in total microbial pollution, a decrease in the number of opportunistic staphylococci and enterobacteria by a factor of 1.5-3.5, and in some cases a decrease in the number of saprophitic bacteria of Sarcina genus. Results of the investigation may be recommended for wide application in order to optimize the indoor conditions of kindergartens.

1 Introduction

The urgency of ecological phytodesign is due to the acute problems related to technogenic and microbiological air pollution and the necessity to care for the health of the population of large cities. Special attention is paid in the world to the indoor air quality in child-care centers because the microbial pollution of indoor air is the reason why various pathologies of upper air passages occupy a stable position in the list of children diseases [1-3]. One of the methods to improve indoor air quality is to use the assortment of tropical plants emitting volatile substances with high antimicrobial activity against opportunistic nicriflora; allergenic, thorny plants and the species with poisonous sap should be excluded. The goal of the present work was to study the effect of the volatile exo-metabolites of plants on the quantitative and qualitative composition of the microflora in the indoor air of kindergartens. This experimental study was aimed at revelation of the interaction of higher

* Corresponding author: ntsybulya@yandex.ru

plants with microorganisms in light of the prophylactic technologies of the improvement of environmental sanitation with respect to the indoor air quality. The microbiological composition of indoor air in kindergartens was monitored taking into account the presence or absence of indoor plants.

2 Materials and methods

Microbiological examination of indoor air was carried out in 12 specialized kindergartens situated in urban regions with different ecological situations. The changes in the microbial composition of microflora in the air (total number of microbes and the fraction of opportunistic microflora) depending on the presence or absence of plants in rooms were determined by measuring the total number of colony-forming units (CFU) in 1 m^3. Automatic sampling of biological aerosol was carried out using a PU-1B sampling device by means of impaction on the solid differential diagnostic nutrient medium GRM agar in five points in a room according to the standard procedure [4]. Air volume for the determination of total microbial number was 100 l, and air volume to determine the number of staphylococci was 250 l. The ratio of the permanent microflora (saprophitic microorganisms of *Micrococcus*, *Sarcina*, *Bacillus* genera) to temporary microflora (opportunistic bacteria of *Staphylococcus* genus, enterobacteria, mold fungi) was determined. Experiments were arranged either to study microbial pollution in the same room before and after installation of plants or in parallel in two identical rooms: plants were placed in one of the rooms (the experimental room), while the other room (reference room) remained without plants till the end of the experiment. Sampling of the air in the rooms of kindergartens was carried out in autumn and winter (November – February) because the highest microbial pollution is observed in Siberia during these months.

The objects of investigation were tropical and subtropical plants which were studied by us previously under laboratory conditions and exhibited high antimicrobial activity with respect to three test objects: *Staphylococcus epidermidis*, *Escherichia coli*, *Candida albicans* [5,6]. We term the plants arranged in a room as a phytomodule, which means the necessary number of phytoncide plants with definite leaf area per unit volume of room, taking into account the phytoecological conditions and types of rooms (the positions of windows with respect to cardinals, air humidity, contingent present in a room). Phytomodules occupy 5-10% of room area.

3 Results and discussion

Most frequently, bacteria dominating in indoor air are gram-positive opportunistic species of *Staphylococcus* genus and mold fungi [4,7]. Indoor air monitoring in several megalopolises in China showed that the qualitative and quantitative composition of microflora in air is unstable and variable depending on the season, time of day, meteorological conditions (wind direction, temperature, air humidity, altitude, insolation) [2]. Some authors from Poland and China stress the differences in the microflora in the indoor air of urban and rural cites and the dependence on the age composition of children in kindergartens, primary and secondary schools [8-10]. However, the problem of choosing the assortment of tropical plants and their effect on the quantitative and qualitative changes in the indoor microbial pollution (иthat is, the composition and total number of microbes) remains urgent.

Results of the quantitative determination of the total number of microbes in the absence of plants and in their presence in rooms are shown in Fig. 1. Total microbial pollution of indoor air after the installation of plants decreased by a factor of 1,5-3,5.

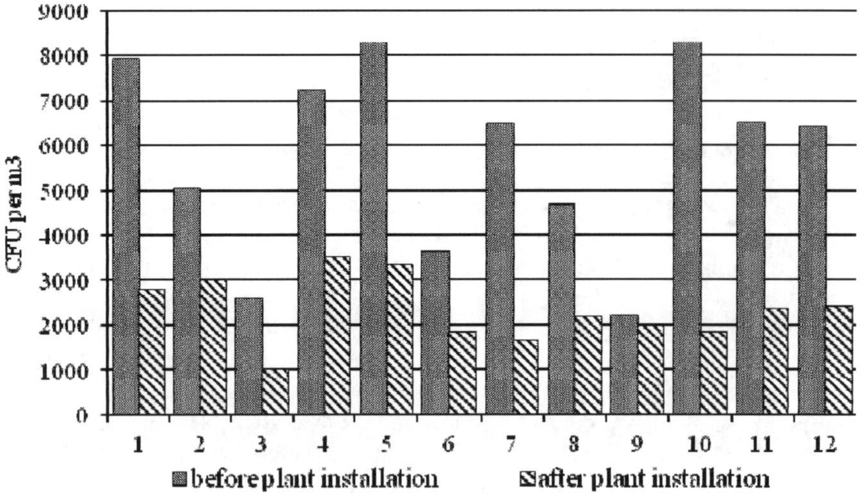

Fig. 1. Results of microbiological examination of some kindergartens in Novosibirsk before and after installation of plants.

To achieve better cogency of the efficiency of sanitizing effect of phytoncide plants, in one of the experiments the plants were installed in the room and then taken away, then again installed, and so on. After plant installation, overriding microbial pollution decreased to the normal level within 84 hours. In the absence of plants, as early as after 4 hours the total number of microbes increased from 5200 to 7500 CFU/m^3 within 52 hours. After the next plant installation, microbial pollution did not exceed the sanitary standard during 118 hours [2].

The analysis of the qualitative changes in the mciroflora allowed us to reveal several features with respect to opportunistic microflora. The fraction of opportunistic microflora was less than 30% in rooms with plants, while in rooms without plants this parameter was 30 to 70% (Fig. 2).

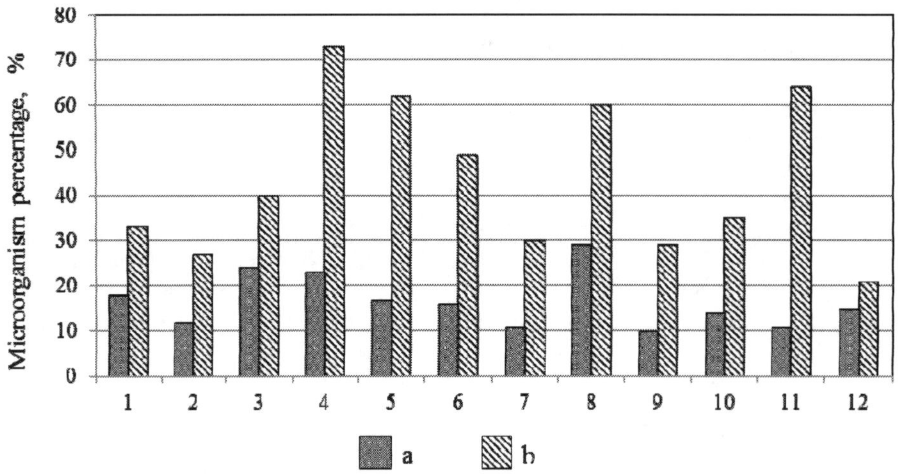

Fig. 2. The percentage of opportunistic (temporary) microflora in the air in comparison with the total number of microbes after (a) and before (b) installation of phytoncide plants.

Volatile substances emitted by plants exhibited activity against opportunistic air microflora. In the absence of plants, the percentage of the bacteria of Staphylococcus genus was 15 to 30% of the total number of microbes, while in the presence of plants it was 5 to 14% (Fig. 3a). After the installation of phytomodules in rooms, the percentage of staphylococci decreased by a factor of 1,5-3. Similar results were obtained also with respect to enterobacteria : 10 to 25% in rooms without plants, 3 to 7% in rooms with plants (Fig. 3b).

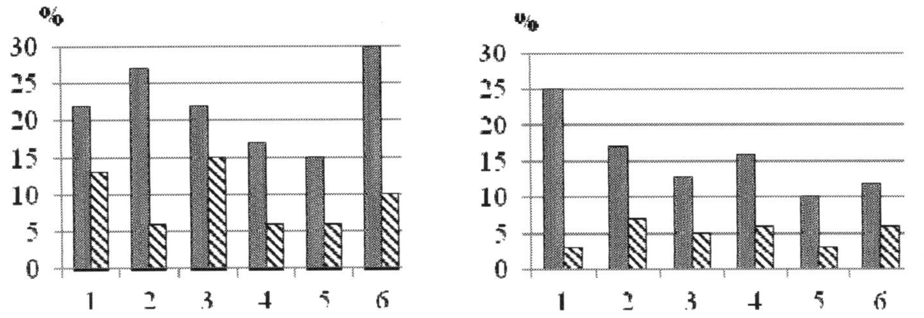

Fig. 3. Percentage of the bacteria of Staphylococcus genus (a) and enterobacteria (b) with respect to the total microbial number in the rooms of preschool institutions before and after installation of phytoncide plants.

The antimicrobial activity of the volatile emissions of *Begonia carolineifolia*, *B. ricinifolia*, *B. boweri*, *B. fischeri*, *Eucharis grandiflora*, *Ficus benjamina*, *F. retusa*, *Nephrolepis exaltata*, *Chlorophytum comosum*, *Kalanchoe pinnata*, *K. blossfeldiana*, *Sansevieria trifasciata*, *Coleus blumei* was tested in two playing rooms (reference – without plants, and experiment – with plants). In this experiment, the leaf area per 100 m^3 of room volume was 3,4 m^2. The reference qualitative analysis of indoor air showed that the percentage of staphylococci with respect to the total number of microbes was 10%, a and the percentage of saprophitic bacteria of sarcina genus was 50 to 80%. After plant installation, a decrease was observed : staphylococci decreased to 5%, sarcinas to 22%, and total number of microbes decreased by a factor of 1,5-3,5 (Fig. 4).

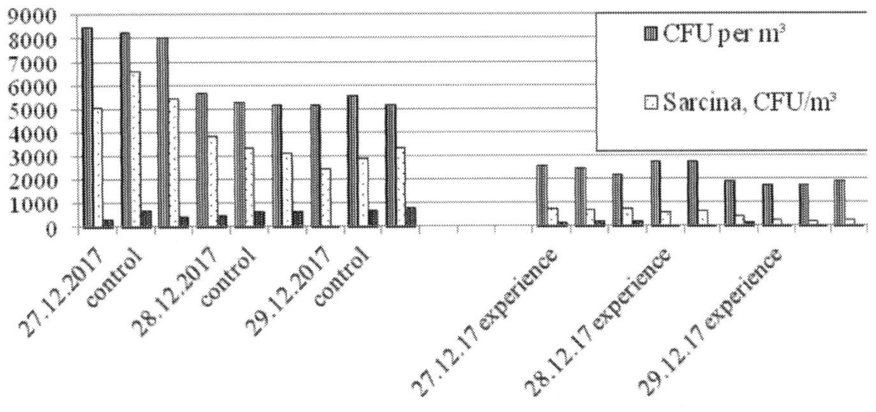

Fig. 4. Changes in the microbial composition in the reference and experimental rooms of a kindergarten after the installation of phytomodule

Experiments aimed at the studies of the antimicrobial activity of volatile plant emissions were carried out directly in rooms, so we determined the leaf area causing pronounced sanitizing effect. The total microbial pollution was studied in 12 preschool institutions ; the leaf area (S, m^2) was varied (Fig. 5). A stable decrease in the total number of microbes to the sanitary standard level in the rooms of kindergartens was achieved with the leaf area 1,5 m^2 to 3,5 m^2 per 100 m^3.

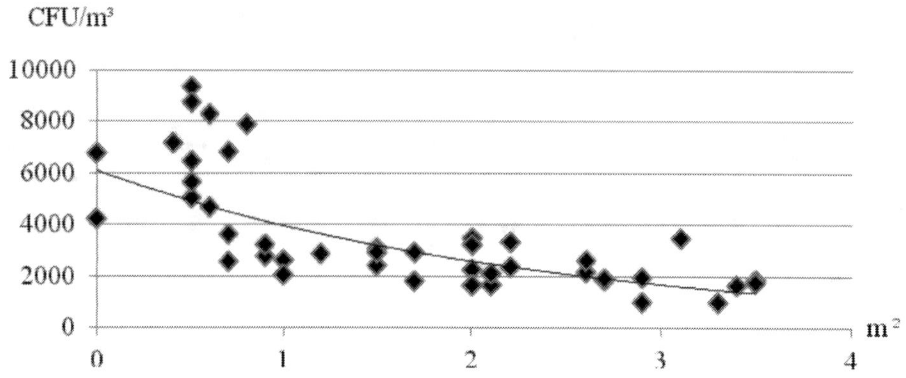

Fig. 5. The degree of microbial pollution of the indoor air in playing rooms of some kindergartens in Novosibirsk depending on the leaf area (m^2) of plants per 100 m^3.

Conclusion

Investigations were carried out directly in the rooms of 12 preschool institutions; the optimal leaf area per unit volume of rooms causing the pronounced antimicrobial effect of the plants was revealed. For the plant leaf area 1.5-3 m^2 per 100 m^3 of room volume, the strong influence of the volatile plant exo-metabolites was observed: a decrease in total microbial pollution, a decrease in the number of opportunistic staphylococci and entorebacteria by a factor of 1.5-3.5, and in some cases a decrease in the number of saprophitic bacteria of sarcina genus. The results on the optimal relations of plant parameters to room volume may be considered as a unique science-intensive technology of indoor air sanitation with the help of antimicrobial phytomodules. These results may be recommended for wide introduction for the purpose of optimization of the conditions in the rooms of preschool institutions.

The work was carried out within the Government assignment for the Central Siberian Botanical Garden SB RAS with partial support from the Russian Foundation for Basic Research under Project No. 17-44-540601. The materials of the bioresource collection of the Central Siberian Botanical Garden SB RAS «Outdoor and Indoor Collection of Living Plants » USU 440534 were used in the investigation.

References

1. L.F. Kaznacheeva, *The problem of frequent children diseases and the routed to the healthy way of life* (Novosibirsk, 1990)

2. L.H. Song, W.M. Song, W. Shi et al., J. Environ. Health **17**, 135-138 (2000)
3. W. Deng, Y. Chai, H. Lin, W.W.M. So, K.W.K. Ho, A.K.Y. Tsui, R.K.S. Wong, Atmos. Environ. **128**, 268-275 (2016)
4. V.V. Pokrovsky, *Clinical Microbiology* (GAETAR, 1999)
5. N.V. Tsybulya, T.D. Fershalova, L.P. Davidovich, *Proceedings of the Samar Scientific Center of the Russian Academy of Sciences* (Samara, 2017)
6. K.-Y. Guan, T.D. Fershalova, N.W. Tsybulia, D.-K Tian, Acta Botanica Yunnanica, **27**, 437-442 (2005)
7. J. Olaitan, O. Adeleke, Internet J. Microbiol., **3**, ispub.com/IJMB/3
8. E. Bragoszevska, A. Mainka, J.S. Pastuszka, K. Lizonczyk, Y.G. Desta, Atmos. **9**, 87 (2018)
9. E. Bragoszevska, A. Mainka, J.S. Pastuszka. *Atmos.* **7,** 142 (2016)
10. G. Zh.-G. Fan, O.Zh.-Y. Uyang, L.-F. Hu, W. X.-K. Ang, M. I. Hong, *Acta Ecologica Sinica* **24**, 315–322 (2004)

Soil cover under Betula pendula Roth. Stands growing on the coal mining spoils in the Kuznetsk region, Russia

Vladimir Ufimtsev[1,*], *Ivan* Belanov[2], and *Olga* Trefilova[3]

[1]The Federal Research Center of Coal and Coal Chemistry of SB RAS, 18, Leningradsky ave., Kemerovo, Russia, 650065
[2] Institute of Soil Science and Agrochemistry of SB RAS, 8/2, Academic Lavrentyev pr., Novosibirsk, Russia, 630090
[3] Sukachev Institute of Forest SB RAS, Federal Research Center "Krasnoyarsk Science Center SB RAS", Academgorodok, 50/28, Krasnoyarsk, 660036

Abstract. The article presents results of the study on the influence of the while birch Betula pendula Roth. stands with varying crown cower (<30, 31-50 and >50%) on the properties of soils developing under these stands on the coal mining spoils in the Kuznetsk hollow (so called Kuzbas region). The studied birch stands were located in two ecotopes, i.e. the northern and southern forest-steppe subzone at Kedrovsk (55°32`33``N 86°04`11``E) and Bungur (54°16`09``N 86°09`00``E) coal mines, respectively. The methodology of the study was based on the concept of plant phytogenic fields. The highest rate of soil organic matter and nutrients accumulation was found under sparse stands in both ecotopes. However, the southern ecotope, as compared to the northern one, was found to have higher organic mater and nutrients content irrespective of the birch stand density.

1 Introduction

The white birch (Betula pendula Roth.) is well known as a pioneer of natural forest revegetation on coal mining spoils in the Kuzbas region [1]. Trees affect soil organic matter composition and spatial distribution, nutrients forms and their content, soil acidity. Trees also alter chemical composition of precipitating water, mediate the rate of detritus decomposition, as well as microbiota composition and biological activity in general [2-5].

The space that experiences the influence of a tree, is called an ecological field (EF) [**Ошибка! Закладка не определена.**, 6]. There are many studies describing the results of investigating the effect of an individual tree or sparse tree stands on soil properties, and the decreasing effect of a tree on soil properties with increasing distance from the tree trunk is well documented. Therefore the EF vertical projection on soil surface is usually differentiated according to the EF effect. Currently there no standardized common approaches for such differentiation, including the number of differentiated subfields.

* Corresponding author: uwy2079@gmail.com

The influence of trees, growing in denser stands with overlapping EF of different trees, on soil properties under them is less studied. The overlapping of individual tree EF is increasing with increasing crown cover, which is manifested by soil cover structure and status [5, 7].

The aim of the study was to compare the influence of white birch stands of different density on the soil properties, developing under the stands during coal mining spoils natural revegetation.

2 Materials and methods

The study was carried out on the coal mining spoils, located near 1) the Kedrovsk coal mine in the northern forest-steppe subzone, 55°32`33``N 86°04`11``E, and 2) Bungur coal mine in the southern forest-steppe subzone, 54°16`09``N 86°09`00``E. The studied soils were technosols [8], or embryozems, i.e. young soils, developing under the canopy of class II birch B. pendula stands aged 28-35 yrs. In many forest inventories the crown cover (CC), defined as percentage of the ground area covered by the vertical projection of crowns, is the main stand density variable. The study sites were chosen according to the birch stand CC: 20-30% (sparse stands), 50-60% (moderately dense stands) and 80-90% (highly dense stands). Within the area of each study site the below-crown, near-crown and beyond-crown concentric circle areas obtained by vertical projections of individual EF on soil surface [6]. In sparse tree stands the external boundary of the below-crown area was determined by vertically projecting external boundary of a tree crown on soil surface, while near-crown circle was determined according to ground plant cover. The total individual EF area was estimated as 4-times of the area of the crown vertical projection on the ground. In moderately and highly dense stand individual EF areas overlapped.

Soil samples were collected from 0-10 cm layer of the predominating soil type within each phytogenic field zone in triplicates. Total soil carbon (STC) and nitrogen (STN) were determined by 2400 Series II CHNS/O Perkin Elmer elemental analyzer. Labile soil phosphorus content was measured colorimetrically in 0.5N acetic acid extracts; exchangeable potassium was measured in 1N ammonium acetate extracts. Soil pH was measured in water extracts potentiometrically by Hanna Instruments pH-meter. Granulometric composition of soil particles was measured by sieving [9], while ≤ 1 mm fraction was determined by laser diffraction using Fritsche Analysette 22.

Plant litter was sampled in thus established zones and then characterized. Projective cover (PC) of vegetation was determined as the percentage of ground area covered by the vertical projection of the respective aboveground phytomass. Soil cover structure was determined by cartography for every CC-specific circle within a phytogenic field.

3 Results and discussion

Individual white birch EF did not have distinct boundaries as judged by their leaf litter, as the latter was found in all 3 concentric circles with its thickness and projective cover in all gradually decreasing with increasing distance from tree trunks.

In highly dense birch stands the horizontal differentiation herbaceous plant cover did not show any differentiation: the projective cover was 8%, being mostly concentrated on elevated microrelief. Overall 32 plant species were recorded, with predominating Melilotus officinalis L. (3% PC), Vicia cracca L. (2% PC), Dactylis glomerata L. (1% PC) and Medicago lupulina L.(1% PC), the rest being represented by singletons. The green phytomass standing stock was estimated as 26±8 g/m2. Organic matter accumulating embryozems (OM-embryozems) accounted for 95%, while initial embryozems were found on just 5% of the study site area.

In moderately dense birch stands the below-crown zones had 30% of the ground plant cover, while the near-crown circles were found to sustain 70%, with beyond-crown circles being absent. In below-crown circles 6 cm-thick birch litter covered 100% of the surface area, with bryophytes occupying just 5%. Overall 9 species were found on this zone, the individual plants being sporadically spread. Herbs were absent. In below-crown circles 1% of the vegetation projective cover was formed by self-disseminated plants of Sorbus sibirica Held. Organic matter accumulating embryozems were found to occupy up to 100% of the area. In near-crown circles 1 cm thick birch litter covered 70%, while bryophytes accounted for 10% of the overall projective cover. Herbs and grasses had 10% of the PC, being represented by 19 species. Such species as Gypsophila perfoliata L. (10 %), Melilotus alba L. (2 %), Centaurea scabiosa L. (2 %) displayed distinct preference for the circle. Standing green phytomass was estimated as 38±7 g/m2. Organic matter accumulating embryozems and initial embryozems accounted for 70% and 30% of the area, respectively.

In sparse stands the below-crown circles accounted for 30% of the area, while near- and beyond-crown zones constituted 30 and 40%, respectively. In below-crown circles the ground was covered with 5-10 cm thick birch litter, while bryophytes could not be found. Overall 8 species were recorded in these stands, with Populus tremula forming 1% of the PC. The growth and development of herbs and grasses under the crowns there was hampered, resulting in low standing green phytomass of just 8±3 g/m2. Soil cover consisted mostly of OM-accumulating embryozems (90% of the area), with the rest of the area occupied by soddy embryozems due to the growth of Poa angustifolia L. with projective cover ca. 15%.

Near-crown circles had a developed bryophyte layer on 50% on the area and high projective cover of 1 cm thick forest litter (50%). Projective cover of herbs reached 30%. Overall 19 species were observed, with 5% of the projective cover formed by each of Melilotus officinalis, Amoria hybrida L. C. Presl., Medicago lupulina, Hippophae rhamnoides L. and 3% formed by each of Centaurea scabiosa and Gypsophila perfoliata, the flag species of the near-crown zone. Standing green phytomass was estimated as 50±11 g/m2. Initial embryozems were entirely absent there, and OM-accumulating embryozems occupied 80-90% of the area. Humus accumulating embryozems were also found on 5% of the area due to abundant legumes.

The beyond-crown circles had still birch litter on 20% of the surface area, the rest of the litter being produces by herbs and grasses. Bryophytes formed 20% of the projective cover. The projective cover of herbs and grasses was 90%. Overall 26 plant species were found. with dominating meadow species such as Dactylis glomerata with its 30 % of the projective cover, Agrostis gigantea L. with 20 %, Achillea millefolium L. with 20 %, Amoria hybrida with 10 % and Phleum pratense L. contributing 10 % into PC. Active growth of grasses and herbes resulted in 7-times increase, as compared to the near-crown circles, in aboveground standing phytomass, reaching 336±11g/m2. Up to 80% of the area was occupied by OM-accumulating embryozems, the rest 20% covered by soddy ones.

The main soil properties for the white birch EF specific circles are shown in Table 1. In sparse stands STC content was the highest, with a minimum in the near-crown circles (4.5-4.8%) and gradually increasing to a maximum at the EF periphery (5.3-6.6%). Such STC spatial distribution is most likely due to the meadow species growth, providing easily decomposable phytomass input with above- and belowground litter and rhizodeposition, which increased with increasing plant performance at some distance from the tree trunks. Irrespective of the stand density, the southern forest-steppe study site had 1.0-1.6% higher STC content, most likely due to 30-50 mm higher yearly precipitation in the area as compared to the northern forest-steppe study site

The STN, generally correlating positively with STC, showed similar pattern of spatial distribution, being maximal in sparse stands and minimal in highly dense stands (Tab.1). Labile P and K contents can be rated as low and medium, respectively. Nevertheless, the tendency of P and K decrease in the below-crown zone is quite clear for all EF-specific circles and both ecotopes studied.

Table 1. Soil properties in the white birch phytogenic field zones, differing in crown cover

Biotope	EF* circle	Dominant soil type	STC, %	pH	K2O, mg/kg	P2O5, mg/kg	STN, %
Sparse stands (crown cover < 30 %)							
Southern Forest-steppe (Kedrovsk coal mine)	near*	OM-accumulating embryozem	4.8	7.49	77	56	0.25
	below	OM-accumulating embryozem	4.6	8.25	53	49	0.29
	beyond	Soddy embryozem	5.3	7.25	112	74	0.31
Northern forest-steppe (Bungur coal mine)	near	OM-accumulating embryozem	4.5	7.01	110	76	0.36
	below	OM-accumulating embryozem	4.9	6.34	93	39	0.34
	beyond	Soddy embryozem	6.6	6.38	132	92	0.35
Moderately dense stands (crown cover < 50%)							
Southern Forest-steppe (Kedrovsk coal mine)	near	OM-accumulating embryozem	3.8	7.25	78	21	0.23
	below	OM-accumulating embryozem	4.1	7.56	66	18	0.21
Northern forest-steppe (Bungur coal mine)	near	OM-accumulating embryozem	4.4	6.17	71	34	0.28
	below	OM-accumulating embryozem	4.8	5.88	59	28	0.35
Highly dense stands (crown cover <90%)							
Southern Forest-steppe (Kedrovsk coal mine)	near	OM-accumulating embryozem	3.4	7.15	92	46	0.15
	below	OM-accumulating embryozem	4.2	7.32	84	25	0.20
Northern forest-steppe (Bungur coal mine)	near	OM-accumulating embryozem	5.1	7.53	87	51	0.25
	below	OM-accumulating embryozem	4.8	8.02	66	62	0.28

*Phytogenic field circles: near-, below- and beyond-crown ones.

Small size particle were found to contribute 18-47% into soil solid particles under sparse stands with 30% CC, the percentage increasing by 3-12% from the near- to beyond-crown circles. The gravel (≥10 mm) percentage, ranging 18-25%, practically did not change in different CC-specific circles. The main shift of 5-12% occurred in the intermediate size fraction from the bigger (10-5 mm) towards the smaller one (3-1 mm). The granulometric composition of the < 1 mm size fraction for all concentric circles can be described as medium loam with physical clay contributing 37.5-41.2%. The fact that granulometric composition was not dependent on the CC-specific circles of the phytogenic field most likely results from the homogeneity of technogenic eluvium. Soil pH also did not vary much, being neutral or slightly alkaline in all study sites.

In other CC circles increased (23-38%) >10 mm particles content and <28% of < 1 mm particles was found. The intermediate fractions (10-5, 5-3 and 3-1 mm) were rather evenly distributed, accounting for 10-20%. The granulometric composition of <1 mm particle can be identified as medium loam.

4 Conclusions

The observed pronounced patchiness of plant cover and soil organic matter accumulation in areas under sparse birch stands suggested that the environment there was most favourable for soil formation and development. Increased soil carbon, soil nitrogen and physical clay content manifested increased quality of soil environment in areas with birch crown cover less than 30% in both northern and southern forest-steppe subzones. The areas with denser crown cover soil were still quite satisfactory in terms of environment quality, albeit somewhat poorer. The obtained results necessitate more critical approach to determine the density of white birch tree stands during reforestation.

References

1. L.P. Barannik, *Bioecological principles of forest recultivation* (Nauka, Novosibirsk, 1988) in Russian
2. T.J.Hokkanen, E.Jorvinen, T. Kuuluvainen, Silva Fennica **29**, 189-203 (1995)
3. A.C.Finzi, C.D.Canham, N.V. Breemen, Ecol. Applications **8**, 447-454 (1998)
4. O.Priha, T.Lehto, A. Smolander, Plant and soil **206**, 191-204 (1999)
5. M.Chodak, M. Niklińska, Biol. Fertil. Soil **46**, 555-566 (2010)
6. W H.u, P.J.H.Sharpe, J.Walker, L.K. Penridge, Ecol. Modell. **29**, 215-243 (1985)
7. S.Kuhlmann-Berenzon, J.Heikkinen, A. Sarkka, Biometrical J. **47**, 517–526 (2005)
8. IUSS Working Group, *WRB, World Reference Base for Soil Resources 2006. International soil classification system for naming soils and creating legends for soil maps* (FAO, Rome, 2006)
9. C.Gromke, B. Ruck, Atmospheric Environment **41**, 3287–3302 (2007)

Winter-hardiness assessment in ground-covering roses by histochemistry

Olga Vasilyeva[1,*], and *Alina* Tzygankova[2]

[1]FSIS "Central Siberian botanical garden SB RAS", Russia
[2]FSEE HPE FSEI "Novosibirsk State Agrarian University", Russia

Abstract. The paper presents the results of the seasonal dynamics of starch and lipids study in annual shoots of rose species and cultivars growing in the urbanized medium of a Siberian megapolis under continental climate conditions. The prospects of prediction the winter hardiness of roses by the rate of starch hydrolysis and lipids storage in pre-winter period are shown. During this period, starch is observed in medullary rays, perimedullary region and pith (centripetal tendency), while primary cortex, phloem and cambium are rich in lipids (centrifugal tendency).

1 Introduction

On the Earth, according to the data of domestic and foreign taxonomists, wild representatives of *Rosa* L. genus number from 350 up to 500, those growing under conditions of both severe high latitudes and mild climates of tropical and subtropical regions. In XIX century, by inter-specific and inter-sectional hybridization the cultivars were created to become a basement for modern groups of garden roses, which are today as many as about 30 by different gardening classifications. Cultivar pool of the world comprises about 30,000 items [1].

Groups of garden roses with high ornamental qualities such as hybrid teas and floribunda are represented by greatest number of cultivars, of which many provide prospects for cultivating even under severe continental climate conditions of West Siberian forest-steppe. Cultivars of those groups bloom on new growth; even in case of complete withering over-ground parts away during wintering, their annual growth is provided by dormant buds burst in the under-ground part of the plant, which survives winter successfully due to deepen planting elaborated by the gardeners of Tomsk.

Impossibility of saving through wintering a significant amount of shoots is the main limitation for using in Siberian landscape gardening most cultivars belonging to ground-covering or climbing and rambling garden groups, as their flowering occurs at the middle parts of biennial shoots. Histochemical analysis of seasonal dynamics of starch, sugars, proteins and lipids in tissues is one of techniques to diagnose winter hardiness in roses [2].

* Corresponding author: vasil.flowers@rambler.ru

Our study targets to assess the level of readiness for wintering in ground-covering roses by dynamics of starch and lipids content in tissues of annual shoots.

2 Materials and Methods

Most of observations and experiments involving ground-covering ornamental plants of various biomorphs have been conducted at the model ground of urban medium on base of Novosibirsk State Agrarian University since 2015. Histochemical techniques were performed in the Laboratory of Introduction of Ornamental Plants of "Central Siberian botanical garden SB RAS". Samples from USU № 440534 (Collection of alive plants growing indoors and outdoors) were used for experiments. Collaborative investigations of ontomorphogenesis in ground-covering plans are being completed according to the project "Revealing the ways of plants' adaptation to contrast dwelling conditions at the levels of populations and organisms", State number AAAA-A17-117012610053-9.

Winter hardiness of species and cultivars was assessed by seasonal dynamics of starch and lipids content in tissues of annual shoots [3]. The following compositions of reagents were taken: 1) for starch storage and hydrolysis analyzing – 2 g of potassium iodide, 0.2 g of crystalline iodine and 100 ml of distilled water; 2) to determine lipids – 0.01 g of Sudan III dye and 20 ml of 96 % ethanol, for further microscopy – 10 ml of glycerol.

Scales for quantitative evaluation of storage substances (Table 1) for roses were elaborated by M.V. Besschetnova [4], the entire content being determined in seasonal dynamics by total score for all the tissues.

Table 1. Evaluation of lipids and starch content in shoot tissues for species and cultivars of roses

Starch grains presence in tissues	Tissue staining for different content of lipids	Score
Occur in single cells	Pale-pink, hardly distinctive	1
Occur in less than 50 % of cells	Light-pink, well distinctive	2
Occur in more than 50 % of cells	Pink, of medium intensity	3
Absent in single cells	Bright-pink	4
Present in all cells	Light-red	5

Main objects for the study were the following:

Rosa maximowiczhiana Regel (a selected form, received from Botanical Garden-Institute FEB RAS and plant collected in theit natural habitat at the south of Primorsky Krai). In wild, it is a deciduous shrub with climbing, rambling or creeping stems up to 5 m long. 5-petalled flowers are white. Once-blooming.

Cultivar 'Excelsa' belonging to the group of hybrids of Rose wichuraiana (Welch, USA, 1908). Shoots are 70-95 cm long. Light-green glossy leaves are 8-9 cm long. Rounded buds open into cupped tightly petalled (60 or more petals) flowers of deep pink color, 3–3.5 cm in diameter, arranged in dense inflorescences with mild fragrance. Once-blooming.

Cultivar 'Fuchsia Meidiland' from the group Shrubs (Meilland, France, 1994). Shoots are 65–85 cm long. Glossy leaves are 8.5–10 cm long. Buds have flattened tops. Cupped-to-flat fuchsia-colored flowers are 5.5–6.5 cm in diameter and compose loose inflorescences. Fragrant is mild. Remontant type of blooming.

3 Results

The efficiency of this technique has earlier been shown by us for species and cultivars of roses relating to different morphotypes [5]. Dynamics of lipids and starch content was being analyzed through the period from may to November. However, that was their localization in rose shoot tissues just in the pre-winter period (late November – early December) to be of most interest.

Comparative analysis of ground-covering roses at tissue and organism levels yielded the results below.

The highest winter hardiness was exhibited in *R. maximowiczhiana* samples. Plagiotropic creeping shoots successfully wintered under the natural snow cover. Orthotropic shoots developing throughout the vegetative period were expelled. Though blooming of the wintered shoots was neither abundant, nor continuous. Seasonal dynamics of lipids and starch content and their localization within tissues in pre-winter period are presented in Fig. 1 and Tables 2, 3.

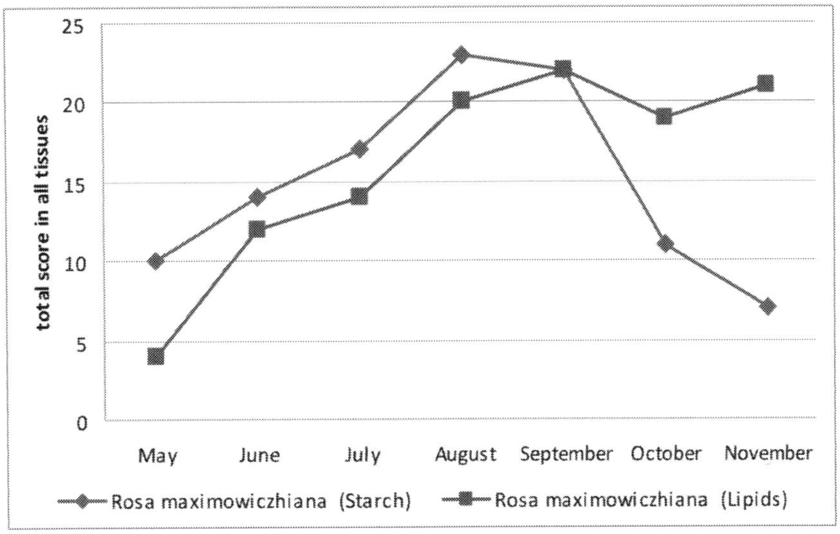

Fig. 1. Seasonal dynamics of lipids and starch content in shoots of winter-hardy ground-covering roses.

Table 2. Starch content in annual rose shoots at their pre-winter stage

Species, cultivar	Primary cortex	Phloem	Cambium	Parenchima of xylem	Medullary rays		Perimedullay region	Pith	Total score
					Uniseriate	Multiseriate			
R. maximowiczhiana	0	0	0	0-1	1	2	1	0-1	4-6
Fuchsia Meidiland	0-1	0	0	1	2	2	1-2	1	8-10
Excelsa	1-2	0	0	1-2	2	2-3	1-2	0-1	11-13

In Tables 2, 3, total score is not calculated automatically for all tissues, but sh varying yearly (that is, there can be 1 point in parenchyma of xylem, or 1 point in uniser medullary rays, but not in both tissues during the same season).

Table 3. Content of lipids in annual rose shoots at their pre-winter stage

Species, cultivar	Primary cortex	Phloem	Cambium	Parenchima of xylem	Medullary rays		Perimedullay region	Pith	Total score
					Uniseriate	Multiseriate			
R. maximowiczhiana	4-5	3-4	2	3	1-3	1	1-3	1-3	20-2
Fuchsia Meidiland	3	2-3	2	2	1	1-2	1	1	13-1
Excelsa	2	1-2	1-2	1-2	1	1	1	0-1	8-10

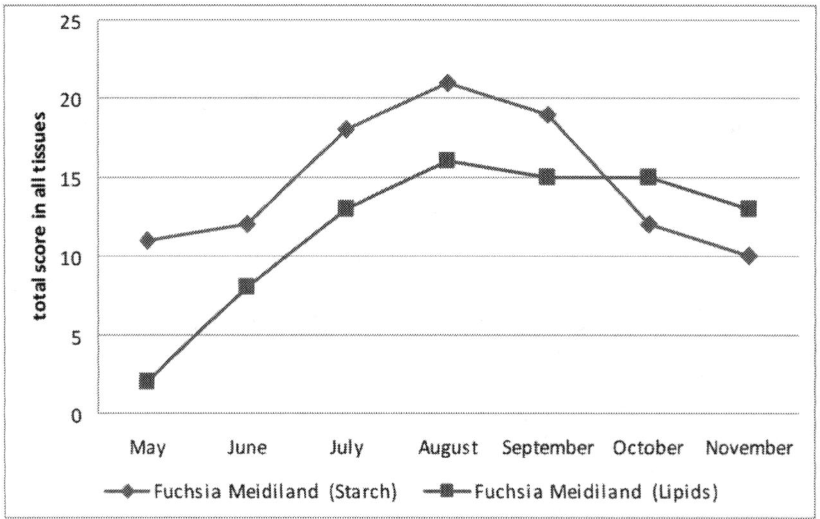

Fig. 2. Seasonal dynamics of lipids and starch content in shoots of ground-covering rose with medium winter hardiness.

Seasonal dynamics of storage substances content reveals that in medium wi hard species *R. maximowiczhiana* outumn starch peak to occurs during the phase of aut foliage coloring (in local winter hard species *R.acicularis* Lindl. and *R. majalis* Herrm maximum is linked to the beginning of abscission). In cultivars with medium and hardiness starck storage maximum is observed at the end of blooming. Actually, varieties under the study do not differ in lipids and starch content in pre-winter perio each other as much as to *R. maximowiczhiana.*

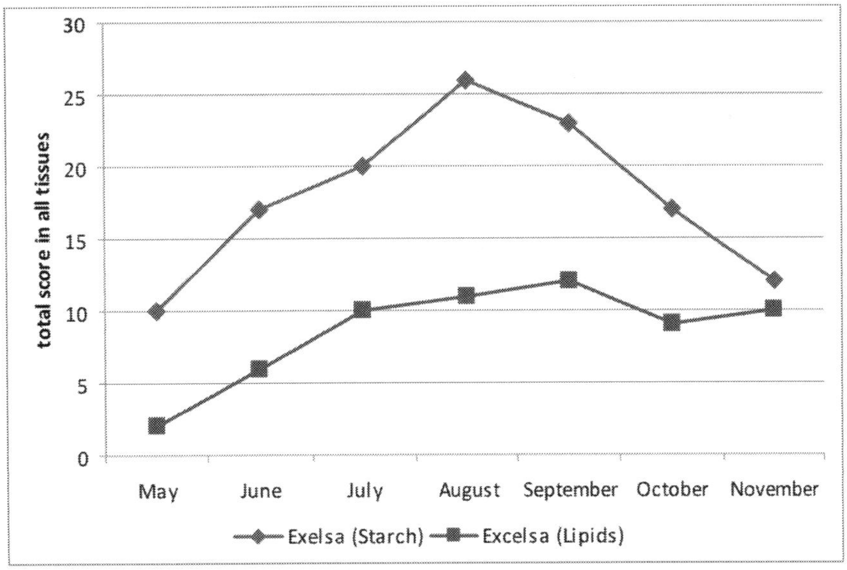

Fig. 3. Seasonal dynamics of lipids and starch content in shoots of ground-covering roses with low winter hardiness

4 Discussion

Nowadays, modern state of investigations in plant hardiness to both high and low temperatures implies application of advanced precision equipment, artificial environment chambers (biotrons); for many studies, model objects, usually herbaceous plants, are used [6–9]. That is of essential importance for profound understanding of acclimatization training mechanisms and cold stress withstanding, as classical histochemical analysis appears insufficient for the purpose. Otherwise, the decided advantage histochemical technique provides is that it enables performing express-assessment of shoot readiness to wintering by quantitative characteristics, in particular starch and lipids localization in certain tissues. The technique does not require expensive reagents, or advanced instrumentation, being widely accessible not only to plant physiologists, but to the ones interested in searching for most resistant gene pools and further breeding and selection.

Seasonal starch transportation within tissues of ground-covering species and cultivars of roses has centripetal pattern; by the end of the vegetative period starch occurs in medullary rays, perimedullary region and pith. On the contrary, lipids are registered in November in primary cortex, phloem and cambium, revealing centrifugal transportation pattern. Shoots of ground-covering roses do not have enough time to complete vegetation, that negatively affecting their winter hardiness.

By and large the group of garden roses under review is not considered as promising for growing under conditions of West Siberian forest-steppe. To expand the possibilities of their cultivation, careful selection of individual cultivars is demanded. An original solution seems to be a combination of two types of ground-covering cultivars – blooming ones and those developing annually the vegetative sphere only.

References

1. M.V. Borzyak, O.Yu. Vasilyeva, State and problems of horticulture in Russia. **2**, 231-235. (1997) (in Russian)
2. O.L. Berezovskaya, T.P. Orekhova, Vestnik FEB RAS, **2**, 129-136. (2011) (in Russian)
3. R.P. Barikina, T.D. Veselova, A.G. Devyatov, Ch.Ch. Djalilova, G.M. Il'ina, N.V. Chubatova, *Handbook of Botanical micro-technology: basics and methods.* (MGU, Moscow, 2004) (in Russian)
4. M.V. Besschetnova, *Roses* (Nauka, Alma-Ata, 1975) (in Russian)
5. O.Yu. Vasilyeva, Horticulture and viticulture **3**, 29-34 (2016) (in Russian)
6. V.I. Avdeev, Vestnik of Orenburg State Pedagogical University. Electronic Scientific Journal **3**, 25-29 (2014)
7. F. Cheng, L. Zhong, N. Zhao, Yi Liu, G. Zhang, Plant Growth Regulation **46**, 87–95 (2005)
8. A. Theocharis, Ch. Cle´ment, E. Barka, Planta **235**, 1091–1105 (2012)
9. C. Partheeban, H. Vijayaraghavan, S. Sowmyapriya, S. Srividhya and D. Vijayalakshmi, Legume Research **40**, 655-659 (2017)

Crown structure of 25-year-old Siberian stone pine grafts from mother trees of different ages

Svetlana Velisevich[1,*], *Alexander* Popov[1,2], and *Sergei* Goroshkevich[1]

[1]Institute of Monitoring of Climatic and Ecological Systems SB RAS, 634055, Tomsk, 10/3 Academichesky ave., Russia
[2]Tomsk State University, 634050 Tomsk, 36 Lenina ave., Russia

Abstract. Inheritance of the age-specific traits of the Siberian stone pine crown structure by vegetative progeny was analyzed. Studies were carried out on 25-year-old grafts onto the same species seedlings as rootstocks. Scions for grafting were sampled from two groups of trees: young-mature (YM) and old growth (OG) trees in the southeast of the West Siberian Plain. The analysis of graft development showed that the YM trees had a significantly higher and thicker trunk, and their crowns were broader and denser due to a larger number of shoots. They multiplied by means of the branching feature due to the predominance of axes 2-3 branching orders. The OG trees formed small and transparent crowns that visually resembled separate branches from the upper parts of the crown of ontogenetically old trees. It was suggested that in the vegetative progeny of trees of a different ontogenetic state, the growth potential was epigenetically inherited.

1 Introduction

Experiments on clonal plantations of conifers have showed that the use of scions from ontogenetically mature donor trees does not allow obtaining vegetative progeny with a high growth rate. Foresters now use young donor trees in breeding for the growth rate [1]. However, the mechanisms underlying the phenomena of memory preservation in separated parts of plants is a subject of scientific discussion [2].

Scientists are interested in Siberian stone pine mainly as a tree producing pine nuts, but the influence of the donor tree's age on the grafts' development remains unknown. The purpose of this work is to study the influence of the donor tree's age on the vegetative growth of the grafts. We also wanted to assess the degree of epigenetic inheritance of age-specific features.

2 Materials and methods

2.1 Object of study

* Corresponding author: s_n_velisevich@mail.ru

Scions for grafts were collected from Siberian stone pine trees in two sites located near Tomsk (southeast of the West Siberian Plain, southern taiga). The first (56°30'N 84°39'E) was a group of 220-year-old old growth (OG) trees. The second (56°20'N 85°08'E) was a group of 30-year-old young-mature (YM) trees. Scions were grafted onto 5-year-old rootstocks. The experiment was carried out at the Kedr Field Station, which is managed by the Institute of Monitoring of Climatic and Ecological Systems, Russian Academy of Sciences (56°13'N 84°51'E). To analyze the crown structure, one average branch was selected in each whorl of the stem, and then the total number of shoots was counted on each branch and multiplied by the number of such branches in the whorl and in each whole of the crown. The apical dominance was calculated from the ratio of the length of the stem to the length of the lateral branch. The crown transparency in the vertical projection was measured using software AutoCAD 2014 (Autodesk, USA).

2.2 Statistical analysis

The whole number of trees of each age group — 26 YM and 32 OG trees — served as a variation series in statistical calculations. The differences between the samples were estimated using the ANOVA F-test. Statistical data processing was carried out using the program Statistica 6.0.

3 Results

Large differences were found between YM and OG grafts (Table 1). YM grafts formed tall and thick trunks. Their crowns were much larger in size and superior to OG trees by the number of shoots and the number of branches per unit length of the trunk. Visually, the OG crowns were similar to individual branches from the crown top of an ontogenetically old tree. In OG trees, the crown was much lighter, with obvious signs of disintegration and isolation of individual large branches in the upper part (Fig. 1). This gave it a candelabra form, which is typical for the old age stage.

Table 1. Crown structure of grafts in different ontogenetic stages.

Trait	YM grafts	OG grafts
	Mean±error	Mean±error
Stem height, m	7.7±0.4*	5.6±0.1
Stem diameter, cm	19.1±1.0*	11.1±0.4
Crown length, m	7.5±0.5*	5.6±0.2
Crown diameter, m	4.5±0.3*	3.2±0.2
Shoot number, pcs.	5428±55.2*	2297±39.1
Branches per unit stem length, pcs./m	712±47.1*	370±56.7
Crown transparency, %	11.6±1.76*	38.8±2.25

Significant differences between age variants for ANOVAs ($P \geq 0.05$) are indicated by asterisks.

In the structure of branching of YM trees there is a greater diversity in the order of branching (Fig. 2). They have a large proportion of axes of the 2nd and 3rd orders, which has a positive effect on the increase in the density of the crown at this age. The crown of OG trees is less diverse in the participation of branches of higher branching orders. They have many axes of the 2nd order and the percentage of skeletal branches of 1st order is relatively high. The part of axes of the 3rd order is much smaller than in YM trees. This type of branching contributes to the formation of an «openwork» crown in OG trees.

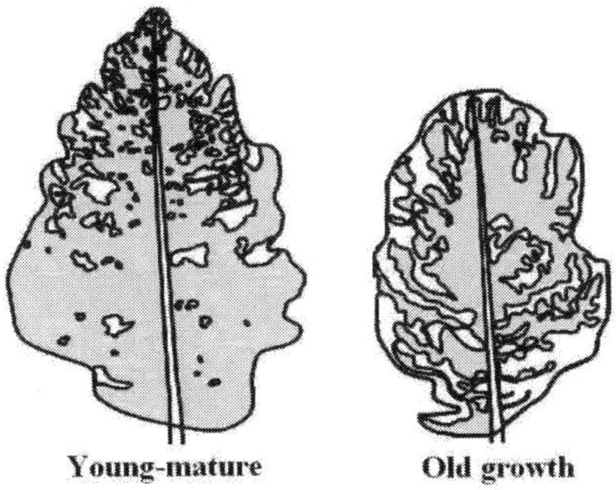

Fig. 1. Crown shape and transparency.

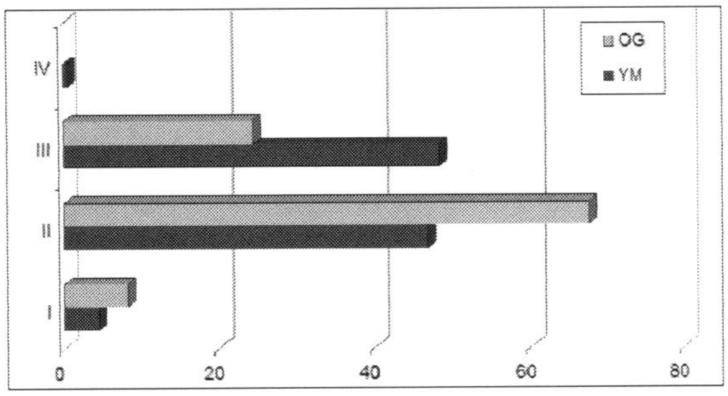

Fig. 2. Branching of YM and OG trees. The abscissa is the branching order, the ordinate is the proportion of shoots of different branching orders, %.

4 Discussion

The problem of ontogeny regulation is far from resolved. The fact that the same traits (simple or complex) can be classified as epigenetically stable or labile complicates the knowledge of the mechanisms of ontogeny regulation. But most researchers agree that the branching pattern exhibits stable epigenetic inheritance [2-5]. Our results are consistent with their conclusions and show the inheritance of an age-specific branch pattern and the shape and density of the crown.

The generally accepted theory of ontogeny is based on the idea of the appearance, strengthening, and disappearance of the capacity for sexual reproduction, which shows itself in a successive change in the virgin, generative, and senile stages. That is, the growth changes in the ontogeny mainly for physiological reasons. This is due to changes in the root–leaf functional correlation, which gradually intensifies as the young tree grows and its vegetative power increases. Each level of this vegetative power corresponds to a new qualitative state of morphogenesis: genes responsible for these qualitatively new phenomena are activated or inactivated. This is assumed to be the essence of ontogenesis; therefore, any of the existing ways of dividing ontogeny into periods are based on qualitative changes in morphogenesis. Following our results, the successive change in the stages of ontogenesis, as the process of realizing the genetic development program, is characterized not by qualitative transformations in morphogenesis, but by growth potential: increase and decrease in vegetative capacity. Therefore, the growth potential is the main trait inherited by vegetative reproduction of trees of different ages.

This work is supported by the Russian Science Foundation under grant no. 18-16-00058.

References

1. J. Kleinschmit, D.K. Khurana, H.D. Gerhold, W.J. Libby, *Clonal Forestry II* (Springer, Berlin, Heidelberg, 1993)
2. M.E. Day, M.S. Greenwood. *Size- and Age-Related Changes in Tree Structure and Function* (Springer, Netherlands, 2011)
3. M.S. Greenwood, M.E. Day, J. Schatz, Tree Physiol. **30**, 459-468 (2010)
4. M.S. Greenwood, C.A. Hopper, K.W. Hutchison, Plant Physiol. **90,** 406-412 (1989)
5. G.A. Ritchie, J.W. Keeley, Tree Physiol. **14**, 1245-1259 (1994)

BIO Web of Conferences **11**, 00044 (2018) https://doi.org/10.1051/bioconf/20181100044
Prospects of Development and Challenges of Modern Botany

Antiviral activity of fungi of the Novosibirsk Region: *Pleurotus ostreatus* and *P. pulmonarius* (Review)

Vyacheslav Vlasenko [1,*], and *Anastasia* Vlasenko

[1] Central Siberian Botanical Garden SB RAS, 630090 Zolotodolinskaya st., 101, Novosibirsk, Russia

Abstract. The article reviews of antiviral properties of two species of fungi from the Novosibirsk Region of the genus *Pleurotus*. Fruit bodies and cultivated mycelium of *Pleurotus ostreatus* and *P. pulmonarius* are a promising source of extracts with polysaccharides, possessing antiviral and anti-cancer properties.

1 Introduction

Pleurotus ostreatus and *P. pulmonarius* are saprotrophic wood-inhabiting basidiomycetes of the *Pleurotaceae* family (figure. 1).

Fig. 1. *Pleurotus ostreatus* (A) and *P. pulmonarius* (B) on various substrates of the Novosibirsk Region.

These mushrooms are edible, their fruit bodies are a valuable dietary product. Oyster mushrooms are cultivated on commercially as a food product. In Novosibirsk Region, in

* Corresponding author: vlasenkomyces@mail.ru

nature, these species are widespread, they grows on many species of woody plants, most often deciduous, but can also grow on coniferous plants.

P. ostreatus and *P. pulmonarius* have medicinal properties, the polysaccharides contained in them exhibit a high antioxidant, antitumor, immunomodulatory, antiviral and other effects [5].

The aim of the study is to review the literature on the antiviral activity of extracts from fruiting bodies and mycelial cultures of *P. ostreatus* and *P. pulmonarius*. We plan to study the dependence of the antiviral activity of the fungi of the genus *Pleurotus* on their strain diversity.

2 Materials and methods

Antiviral activity has been identified in many agaricoid and polyporoid fungi. In Russia, the antiviral activity of wood-inhabiting basidiomycetes of Western Siberia was studied by us together with the employees of the State Research Center of Virology and Biotechnology VECTOR (Koltsovo, Novosibirsk Region) [8, 15, 17].

3 Results and Discussion

The antiviral properties of fungi began to be studied in the second half of the 20th century. For the first time in 1979, data were obtained on the antiviral activity of the extract from *Lentinula edodes* [13]. In 1969, the antitumor activity of a polysaccharide extracted from the fruit body of *P. ostreatus* [20].

Polysaccharides of oyster mushrooms were first isolated from *P. ostreatus*. It was an insoluble polysaccharide Pleuran [7], commercial name Imunoglukan®. It possesses immunomodulating properties that make the body more resistant to infections and cancer [10] and has other properties. It was found that polysaccharides possess antitumor activity [1].

The antiviral properties of aqueous and methanol extracts of *P. ostreatus* were evaluated against herpes simplex virus type 1 (HSV-1). Antiviral activity correlates with the beta-glucans present in the polysaccharide fraction, which showed higher antiviral activity than aqueous extracts [12].

Water extracts with polysaccharides *P. ostreatus* and *P. pulmonarius* were active against the herpes simplex virus of the second type. The antiviral activity of water extracts of fungi is associated with the presence of polysaccharides and increases with increasing concentration of polysaccharides [2].

Extracts with polysaccharides obtained from water extracts from the genus *Pleurotus* completely suppress the infectious activity of at least 1000 TCID50 of the West Nile virus [11].

Aquatic extracts of *P. pulmonarius* fruiting bodies showed high activity against influenza viruses (6.06 ± 0.18 lg for H5N1 and 5.73 ± 0.14 lg for H3N2) [6].

The antiviral activity of the in vitro mycelium of higher fungi against influenza A virus (serotype H1N1) and herpes simplex virus type 2 (HSV-2), strain BH was investigated. *P. ostreatus* inhibited the propagation of the influenza A / FM / 1/47 (H1N1) virus strain in MDCK cells that reduce the infectious titer by 2.0-6.0 lg ID50. It was determined that *P. ostreatus* is effective against HSH-2 strain BH in RK-13 cells with similar levels of inhibition as for influenza [9].

Ubiquitin-like glycoprotein was isolated from *P. ostreatus*, which inhibited the development of the human immunodeficiency virus [18].

It was found that the purified lectins of *P. ostreatus* as an adjuvant (1 mg / ml) also enhance the immunogenicity of the hepatitis B DNA vaccine [3].

4 Conclusions

Many polysaccharides exhibiting antiviral effects also possess antitumor properties [16, 19]. Data on the antitumor and antiviral activity of the same species of fungi indicate a correlation between these properties [14]. The role of extracts obtained from basidiomycetes can only increase, because in addition to the prevention and treatment of viral infections, polysaccharides and other metabolites of fungi will contribute to the prevention of types of cancer that have a viral etiology [4].

The reported research was funded by Russian Foundation for Basic Research and the government of the Novosibirsk Region of the Russian Federation, grant № 18-44-543018.

References

1. P. Bobek, E. Ginter, L. Kuniak, J. Babala, M. Jurcovicova, L. Ozdín, J. Cerven, Nutritio **7** , 105-108 (1991)
2. G. Chihara, J. Hamuro, Y.Y. Maeda, Y. Arai, F. Fukuoka, Cancer Res. **30**, 2776-2781 (1970)
3. W. Gao, Y. Sun, S. Chen, J. Zhang, J. Kang, Y. Wang, H. Wang, G. Xia, Q. Liu, Y. Kang, Vaccine **31**, 2273-2280 (2013)
4. V.E. Gurtsevich, The Russian Therapeutic Journal **7**, 8 (2008) (in Russian)
5. I.A. Gorbunova, V.A. Vlasenko, T.V. Teplyakova, T.A. Kosogova, I.N. Mikhaylovskaya, Coniferous of boreal zone **26**, 12-21 (2009) (in Russian)
6. A.S. Kabanov, T.A. Kosogova, L.N. Shishkina, T.V. Teplyakova, M.O. Skarnovich, N. A. Mazurkova, L.I. Puchkova, E.M. Malkova, E.A. Stavskiy, I.G Drozdov, Journal of Microbiology, Epidemiology and Immunology **1**, 40-43 (2011) (in Russian)
7. S. Karacsonyi, L. Kuniak, Carbohydrate Polymers **24**, 107-111 (1994)
8. N.E. Kostina, Zh.B. Ibragimova, M.A. Protsenko, E.V. Makarevich, M.A. Skarnovich, E.I. Philippova, I.A. Gorbunova, V.A. Vlasenko, G.P. Troschkova, N.A. Mazurkova, L.N. Schischkina, Modern problems of science and education **3** (2013) (in Russian)
9. T. Krupodorova, S. Rybalko, V. Barshteyn, Virol Sin. **29**, 284-190 (2014)
10. N.R. Di Luzio, Springer Seminars in Immunopathology **8**, 387-400 (1985)
11. I.A. Razumov, T.A. Kosogova, E.I. Kazachinskaya, L.I. Puchkova, N.S. Shcherbakova, I.A. Gorbunova, I.N. Mikhailovskaya, V.B. Loktev, T.V. Teplyakova, Antibiotics and Chemotherapy **55**, 14-18. (2010) (in Russian)
12. S. Santoyo, A.C. Ramírez-Anguiano, L. Aldars-García, G. Reglero, C. Soler-Rivas, Journal of food and nutrition research **51**, 225-235 (2012)
13. M. Takehara, K. Kuida, K. Mori, Arch Virol. **59**, 269-374, (1979)
14. T.V. Teplyakova, T.A. Kosogova, Higher Fungi of Western Siberia as Promising Objects for Drug Production Biotechnology, **298** (2014) (in Russian)
15. T.V. Teplyakova, N.V. Psurtseva, T.A. Kosogova, N.A. Mazurkova, V.A. Khanin, V.A. Vlasenko, Int. J. of Med. Mushrooms **14**,37-45 (2012)

16. S. Tochikura, Med. Microbiol. Immunol. **177**, 235-244 (1988)
17. V.A. Vlasenko, T.V. Teplyakova, N.A. Mazurkova, T.A. Kosogova, A.V. Bardaschova, N.V. Psurtseva, Bulletin of the Altai State Agrarian University **90**, 29-31 (2012) (in Russian)
18. H.X. Wang, T.B. Ng, Life Science **68**, 2151-2158 (2001)
19. S.P. Wasser, A.L. Weis, Int. J. of Med. Mushrooms **1**, 31-62, (1999)
20. T. Watanabe, Nippon Rinsho **27**, 1759-1761 (1969)

Thematic interpretation of the Northern Dvina River floodplain meadow ecosystem using remote sensing

Alexey Volkov*, and *Tatyana* Parinova

Northern (Arctic) Federal University named after M.V. Lomonosov. Higher School of Natural Sciences and Technologies. Department of biology, ecology and biotechnology. Arkhangelsk, Russian Federation

Abstract. In this work we represent the possibility of using satellite images of medium spatial resolution for thematic interpretation of meadow ecosystems located in the flood plain of the Northern Dvina river. We have used multidimensional scaling to study the differences in spectral characteristics. Classes such as arable land, water, mineral soils, forests and swamps are most well-classified. However, the meadow ecosystems overlap with reed beds according to spectral characteristics. Therefore, for more accurate identification of meadows, it is recommended to use different vegetative indices, for example NDVI and Cl Green. As a result of the research, a map of the distribution of meadow ecosystems was created using GIS.

1 Introduction

One of the traditional tasks for geobotany is the mapping of vegetation, including meadow, which is the object of our many years of research. Large-scale data on areas of meadow ecosystems in the Russian Federation refer mainly to the 1960-1980th years and at the present time do not adequately reflect the current situation [1]. Meadows in the floodplain are of high natural and economic value. Among other ecosystems, they are particularly dynamic and undergo significant changes over time. Their areas, the degree of moistening, the composition of the herbage are constantly changing. Therefore, it is important to have a reliable and affordable way of remote meadow registration and evaluation of their areas.

The aim of this study is to identify criteria that allow deciphering the meadow communities and create a map of the floodplain of the Northern Dvina River. According to this, the following tasks are set: to determine the spectral characteristics of meadows from satellite images, to compare them with other classes of objects, to identify differences, to establish the relationship of spectral characteristics to the values of vegetation indices, and to conduct thematic interpretation with the subsequent compilation of a map of the flood plain of the Northern Dvina River.

* Corresponding author: a.g.volkov@narfu.ru

2 Materials and Methods

The meadows in the floodplain of the Severnaya Dvina River representing grassy communities with more or less contiguous herbage and formed mainly by long-term mesophilous plants [2] were chosen as the research subject.

We have used several data sources for thematic interpretation of meadow ecosystems.

Firstly, during field expeditions in 2011-2017, in the floodplain of the Severnaya Dvina River, we have detected meadows and located their coordinates. Later they were added on the satellite image and used as reference sites.

Secondly, we used satellite images from the Sentinel-2 (07/24/2017) with a spatial resolution of 10, 20 and 60 m / pixel, depending on the channel. All spectral channels from MSI were included in the analysis.

Thirdly, to improve the accuracy of the classification, we used satellite-derived data, such as the vegetation index. First of all, this is NDVI (1), which reflects the relative amount of photosynthetically active biomass (2). And also Cl Green (2) – the index of photosynthetic activity of the vegetation cover, used in assessing the chlorophyll a and b content in plant leaves [3, 4].

$$NDVI = (NIR - RED) / (NIR + RED) \tag{1}$$

$$Cl\ Green = NIR / Green - 1 \tag{2}$$

Where NIR – near infrared channel; RED – red channel; GREEN – green channel.

Using ArcGIS, we created a vector polygonal layer with known natural objects: meadows, arable land, forests, reed beds, marshes, water and anthropogenic. For each class we created 1000 random points and extract values from all data sets. Then we exported the data to csv file and analysed in R. Multidimensional scaling with Bray-Curtis distance used for searching the differences between classes. Multidimensional scaling makes it possible to visualize multidimensional feature data in a two-dimensional space of coordinates based on the distance matrix. In addition, it is possible to find a correlation with the factors, determine their influence on the differentiation of objects and display them in the form of vectors on the graph.

The interpretation was also carried out in R using the "random forest" method [5] which is an ensemble learning method for classification that operates by constructing a multitude of decision trees at training time and outputting the class that is the mode.

3 Results

The above classification of the satellite image revealed significant differences in the spectral characteristics of various deciphered objects. As expected, classes of arable land, water, mineral soils (such as sandbanks in the delta of the river), forests and marshes are best separated. At the same time, it is quite difficult to distinguish between meadow ecosystems and the so-called reed (Figure 1A), which are heavily waterlogged areas with periodic seasonal underflooding, where reed *Phragmites australis* (Cav.) Trin. ex Steud. L. predominates.

Vegetative indices can be used for more accurate interpretation of meadow ecosystems and reed thickets. In our work we used NDVI and Cl Green. In R we calculated the correlation coefficients between these indices and multidimensional scaling axes, containing compressed information on the spectral characteristics of the objects. Meadow communities as a whole have higher values of the applied indices (Figure 1B), which corresponds to more phytomass and chlorophyll content in meadow grass plants than in coarse phytomass of reeds.

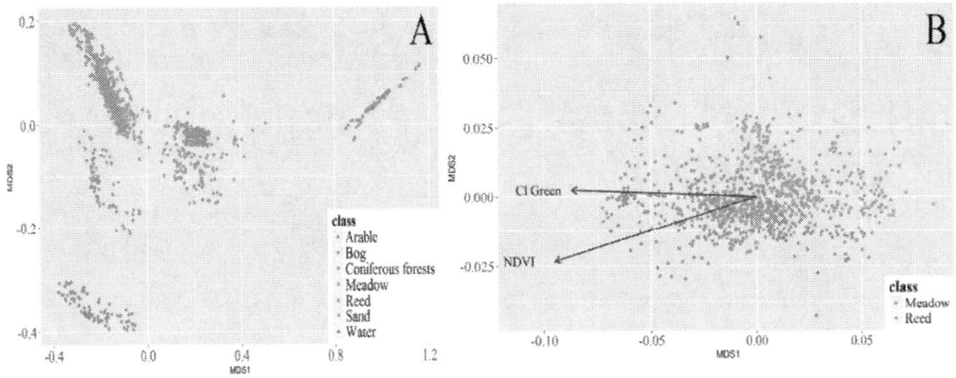

Fig. 1. The results of multidimensional scaling: A – for all classes of objects, B – for meadows and reed thickets with superposition of vectors of vegetative indices.

The obtained information on spectral characteristics and vegetation indices was used to decode the entire satellite image. Based on the results of the thematic interpretation, we have created a vector map of the meadow ecosystems of the floodplain of the Northern Dvina River.

The purpose of compiling any map is to present, at a certain scale, the objects of observation for their recording, as well as to study the patterns of their spread in connection with various natural factors. As a result of the study, spectral characteristics of meadow ecosystems located in the floodplain of the Northern Dvina River were determined. It is revealed that the meadows as a whole are well interpreted with respect to other classes of objects, such as forest, bog, arable and water. However, for the separation of meadows from classes with similar vegetation, it is necessary to use vegetative indices.

This project is supported by grant from the Russian Foundation for Basic Research (RFBR) and the government of Arkhangelsk region № 17-44-290111 (project manager: Prof. Elena Nakvasina).

References

1. T.K. Yurkovskaya, *Actual problems of geobotany*, 43-71 (Petrozavodsk, 2007)
2. B.M. Mirkin, L.M. Gareyeva, Bulletin of the Moscow Society of Naturalists **83**, 38-50 (1978)
3. J. Rouse, R. Haas, J. Schell, D. Deering, 3rd ERTS Symposium, NASA SP-351, **1**, 309-317 (1973)
4. G. Dall'Olmo, A.A. Gitelson, Applied Optics **44**, 412-422 (2005)
5. T. Hastie, R. Tibshirani, J. Friedman, *Random Forests. The Elements of Statistical Learning: Data Mining, Inference, and Prediction* (Springer-Verlag, 2009)

Variation of nuclear DNA content in seeds of *Nitraria schoberi* L.

Maria Voronkova[1,*], *Evgeniy* Banaev[1], *Maria* Tomoshevich[1], and *Taigana* Ak-Lama[1]

[1]Central Siberian Botanical Garden of the Siberian Branch of the Russian Academy of Sciences, 630090, Novosibirsk, Russian Federation.

Abstract. For the first time are search of nuclear DNA relative content in *Nitraria schoberi* L. seeds from 15 natural populations of Siberia, the Crimea, Kazakhstan, and Tajikistan has been conducted by the method of flow cytometry. High intra- and interpopulational variations of nuclear DNA content - 2,93-3,39 pg, at average value - 3,22±0,108 pg is revealed.

1 Introduction

The assessment of intraspecific variation of a genome size helps cognition of evolution pattern and species adaptation, way soft heir divergence, and solving systematic issues in complex taxonomic groups [1, 2]. *Nitraria* L. species are typical halophytes, their populations are often small and isolated since they are restricted to intrazonal communities. *Nitraria schoberi* L. Is the most polymorphic species, it is characterized by mixoploidy and variability of chromosome numbers which may be one of the factors of plant adaptation to habitat extreme conditions [3, 4].

The aim of the work was study of relative content of nuclear DNA in *N. schoberi* in different habitats within the natural range.

2 Materials and Methods

2.1 Plant materials

The content of nuclear DNA was determined in *N. schoberi* dormant seeds. As an internal standard were used fresh leaves of *Raphanus sativus* L. ′Saxa′[32] (2C DNA content = 1,11 pg) seeds of which were received from the Centre of Plant Structural and Functional Genomic, Institute of Experimental Botany, Academy of Sciences of the Czech Republic, Olomouc[5].

Research was conducted in 15 natural populations of Siberia, the Crimea, Kazakhstan, and Tajikistan (Figure 1). For the analysis of an endogenic form of variation, 4 model plants were selected in three populations (№ 5, 11, 15), 6 seeds from each plant were analyzed. For the analysis of intrapopulational variation, seeds were collected not less than

*Corresponding author: bmc_87@mail.ru

from 10 plants in each of the populations. Seeds were collected by the authors of the article in 2009-2017, they are stored in CSBG SB RAS.

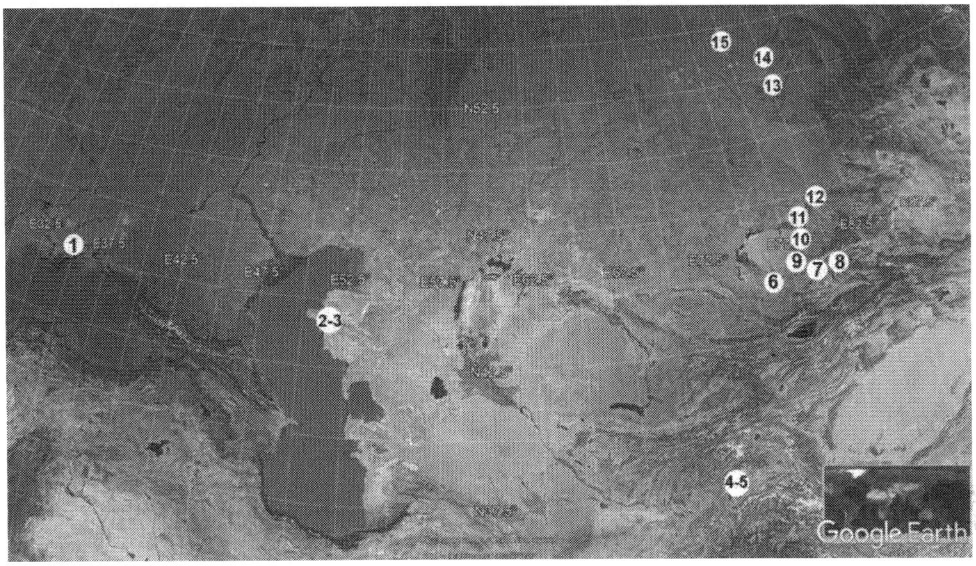

Fig. 1. Collection sites of the plant material. 1-15 - numbers of populations.

2.2 Flow Cytometry

All measurements were performed in Central Siberian Botanical Garden, SB RAS (Novosibirsk). DNA content of *N. schoberi* plants was determined by the method of flow cytometry using staining of isolated nuclei with propidium iodide (PI). The analysis was performed on Cy Flow Space (Germany, Sysmex Partec) with a laser radiation source of 532 nm. The procedure of the work in detail was given by us earlier [6].

2.3 Statistical analysis

Basic statistics (mean, standard deviation, genome size variation) were calculated in R [7].

3 Results

The analysis of endogenic variations showed sufficiently high variability of relative DNA content in *N. schoberi* seeds. The greatest stability was revealed in model plants of population №15 growing in the vicinities of Bagan Lake, Novosibirsk Oblast - 3,28-3,39 pg (genome size variation (V) - 0,24-0,39%). In the populations of Tajikistan (№5) and Kazakhstan (№11) DNA content in seeds of one plant changed within 3,09-3,32 pg and 3,18-3,34 pg, respectively, V = 0,25-1,44 %. The data obtained on variation of DNA content in seeds collected from one plant correlated with intrapopulational variability of DNA content (Table). So minimum intrapopulational variation was found in population № 6 (Kazakhstan) - 0,39 % and maximum one - in population № 5 near the Pyandzh River (Tajikistan) - 1,56 %. In this regard it should be noted that seeds collected from one bush may have a male hametophyte from the other plant, and consequently the results obtained may be manifestation of intrapopulational variation. Interpopulational variation of DNA content in *N. schoberi* seeds amounted to 3,35%.

Table 1. Intrapopulational variation of the relative DNA content in *N. schoberi* seeds

№ of population	Quantity of studied samples	2C DNA content (pg) average value	SD*	Min	Max	V**, %
1	10	3,09	0,017	3,06	3,11	0,55
2	10	3,23	0,015	3,21	3,25	0,46
3	10	3,20	0,020	3,16	3,23	0,63
4	10	3,25	0,016	3,22	3,26	0,49
5	36	3,21	0,050	3,09	3,32	1,56
6	10	3,07	0,012	3,05	3,09	0,39
7	10	3,32	0,015	3,30	3,34	0,45
8	10	3,27	0,027	3,23	3,30	0,83
9	10	3,17	0,019	3,16	3,21	0,60
10	10	2,97	0,013	2,95	2,99	0,44
11	24	3,25	0,040	3,18	3,34	1,23
12	10	2,98	0,031	2,93	3,02	1,04
13	10	3,21	0,032	3,16	3,27	1,00
14	10	3,22	0,035	3,16	3,28	1,09
15	30	3,34	0,036	3,28	3,39	1,08
Average value	210	3,22	0,108	2,93	3,39	3,35

*SD – standard deviation, **V – genome size variation

4 Discussion

Occurrence of intraspecific variation of the genome size in various taxa was presented in a number of works, however, the results obtained were often refuted by replicate experiments for various reasons - the international interior standards were not used, availability of secondary metabolites preventing staining was ignored [2]. Such reports led to scepticism of reliability of any data on intraspecific variation of DNA content. Study of intraspecific genome size variation became again popular when the technique of experiments was improved and flow cytometry was widely adoptedas an analytical method [8-10].

It is assumed that changes in the genome size may be related to species divergence, different environmental conditions, development stages touched on various populations or certain plants [11-15]. It should be noted that variation of 2C values determined by heterochromatin polymorphism, B-chromosomes, polyploidy or hybridization is well understood and is not exceptional for the species of angiosperms [2].

Our research showed that nuclear DNA content in *N. schoberi* seeds varied on the same level within both one plant (from 1,03 to 1,08 times) and a population (from 1,01 to 1,08 times). Extreme variants within the whole studied range differed in DNA content by 1,16 times. We did not reveal here with dependence of DNA content in plants on geographical position of populations or habitat conditions. The most likely explanation of variability of DNA content in *N. schoberi* seeds is a widespread aneuploidy in the genus *Nitraria* [16]. Karyological study showed that there was no certain number of chromosomes in *N. schoberi* cells, only the most frequently encountered 2n=4x=48 [3, 4, 17]. In a number of works was noted influence of aneuploidy on the genome size and shown that the method of cytometry was sensitive and informative enough for detection of that phenomenon [18, 19]. Further more thorough study of *Nitraria* species, including karyology of *N. schoberi* plants in different populations, and comparison of the data obtained with the results of cytometry are needed.

The study was supported in part by the Russian Foundation for Basic Research (grant 16-04-00631 to E.V. Banaev).

References

1. P. Smarda, P. Bures, Preslia –Praha **82,** 41-61 (2010)
2. J. Greilhuber, Ann. Bot. **95**, 91-98 (2005)
3. E. N. Muratova, O. V. Goryachkina, E. V. Banaev, Turczaninowia **16,** 50-54 (2013). In Russ.
4. I.Tarnavshi, Bul. Gradini Bot. al Muz. Bot. Univ. din Cluj. **28**, 1-130 (1948)
5. J. Doležel, S. Sgorbati, S. Lucretti, Physiol. Plant **85**, 625-631 (1992)
6. E. V. Banaev, M. A.Tomoshevich, M. S. Voronkova, Bot. Pacifica **7**, 89-92 (2018)
7. R Development Core Team. R: a language and environment for statistical computing. (Vienna, 2011) http://www.R-project.org
8. J. Doležel, J. Bartoš, Ann. Bot. **95**, 99-110 (2005)
9. J. Doležel, J. Greilhuber, J. Suda, *Flow cytometry with plant cells: analysis of genes, chromosomes and genomes* (Wiley, Weinheim, 2007)
10. J. Greilhuber, Ann. Bot. **101**, 791-804 (2008)
11. H. J. Price, K. L. Chambers, K. Bachmann, Bot. Gaz. **142,** 415-426 (1981)
12. A. Cavallini, L. Natali, Caryologia **44**, 93-107 (1991)
13. L. Natali, A. Cavallini, G. Cionini, O. Sassoli, P. G. Cionini, M. Durante, Theor. Appl. Genet. **85**, 506-512 (1993)
14. H. J. Price, J. S. Johnston, Proc. Natl. Acad.Sci. USA **93**, 11264-11267 (1996)
15. J. Loureiro, E. Rodriguez, J. Doležel, C. Santos, Ann. Bot. **98**, 515-527 (2006)
16. E. N. Muratova, O. V. Kvitko, E. V. Banaev, J.-Q. Zhang, G.-Zh. Wang. Bot J **96**, 108-115 (2011). In Russ.
17. G. Reese. Flora. **146**, 478-487 (1958)
18. M. Pfosser, A. Amon, T. Lelley, E. Heberle-Bors, Cytometry **21**, 387-393 (1995)
19. N. Roux, A. Toloza, Z. Radecki, F. J. Zapata-Arias, J. Doležel, Plant Cell Reps. **21**, 483-490 (2003)

BIO Web of Conferences **11**, 00047 (2018) https://doi.org/10.1051/bioconf/20181100047
Prospects of Development and Challenges of Modern Botany

Phenolic compounds of *Bistorta alopecuroides* (Turcz. ex Meissn.) Kom. (Polygonaceae) from different populations of Siberia and the Far East

Maria Voronkova[1,*], and *Galina* Vysochina[1]

[1]Central Siberian Botanical Garden of the Siberian Branch of the Russian Academy of Sciences, 630090, Novosibirsk, Russian Federation.

Abstract. The article considers research of the phenolic compounds composition and content in leaves of *Bistorta alopecuroides (Turcz. Ex Meissn.) Kom.* from different populations using HPLC methods. It shows that *B. alopecuroides* component composition is quite variable and depends on many environmental factors, such as microclimate, altitude, technogenic pollution. Isovitixin, rutin and flavonolglycoside (t_R = 9.2 min; UV 273, 355 nm) compounds are found in *B. alopecuroides* leaves of all studied populations. *B. alopecuroides* growing in high mountains and under technogenic pollution conditions have the highest content of phenolic compounds, which confirms their adaptive role in plant metabolism. The total content of phenolic compounds in leaves of *B. alopecuroides* plants reaches 20.82mg/g.

1 Introduction

The genus *Bistorta* (L.) Scop. species contain a variety of phenolic compounds - anthocyanins, catechins, flavones, flavonols, phenol carboxylic acids. It is well known their anti-inflammatory, antitumoral and antibacterial properties [1]. Biological activity of *B. officinalis* Delarbre and *B. vivipara* (L.) Delarbre [2], two widely spread in Eurasia species, is being active studied. Other species of the *Bistorta* genus, including *B. alopecuroides*, have been investigated insufficiently.

Species of the *Bistorta* genus are high polymorphic, their morphological races are poorly isolated and have a number of transitional forms. Studying phenolic compounds is urgent to reveal important biochemical signs, to clarify the taxonomic affiliation, and to identify perspective plants for medicine use.

The article objective is to investigate the phenolic compounds composition and content in leaves of *B. alopecuroides* from different populations of Siberia and the Far East.

2 Materials and Methods

* Corresponding author: bmc_87@mail.ru

The extracts of *B. alopecuroides* leaves of plants growing in Asian Russia were studied by HPLC. Table 1 shows sites of the samples collecting. All the plants were gathered during the flowering phase.

Table 1. *B. alopecuroides* samples

№	Collection place
1	The Republic of Buryatia, Tunka District, right bank of the Margasan River inflow, mixed forest canopy
2	Transbaikal Region, Duldurgin District, Ara-Ilya settlement surroundings, "Alkhanai" National Park cordon, Dybyksa tract
3	Amur Region, Oktyabrsky District, Erkovets coal open cast, reed-grass meadow
4	Amur Region, Mazanov district, Lebedikha village surroundings
5	The Republic of Buryatia, Tunka District, Nilovka settlement surroundings, Nilova Pustyn' resort, the Ikhe-Ugun River bank, pine-birch-willow forest, 919 m above s.l.
6	Amur Region, Blagoveshchensk city, 11-12 km of Novo-Troitsk autoroad, dry meadow
7	Amur Region, Blagoveshchensk District, 47 km of Ignatievsk autoroad, Mikhailov Pillars
8	Irkutsk Region, the Zunduk River riht bank, Lake Baikal shore, paludified peddle stones
9	Amur Region, Blagoveshchensk city, 10-11 km of Novo-Troitsk autoroad
10	The Republic of Buryatia, Zakamensk District, Bayangola settlement surroundings, the Sangina River right bank, deciduous forest (burnt), 1140 m above s.l.

The analysis of glycosides in water-ethanol extracts was carried out by high-performance liquid chromatography on the analytical HPLC system consisting of Agilent 1200 liquid chromatograph with a diode array detector and ChemStation, a system for collecting and processing chromatographic data. The separation was done on a Zorbax SB-C18 column 4.6 x 150 mm in size, with a particle diameter of 5 μm using a gradient elution mode. To separate glycosides in the mobile phase, the methanol content in the aqueous solution of orthophosphoric acid (0.1%) varied from 32 to 33% in 27 minutes; from 33 to 46% since 27 to 38 minutes; from 46 to 56% since 38 to 50 minutes. The volume of the injected sample was 5 μl, the eluent flow rate - 1 ml/min, under column temperature 26° C. [3]. The quantitative detection of individual compounds in eluates was carried with an external standard [4]. The content of phenolic compounds was calculated as milligram of quercetin equivalents per gram dry weight of sample (mg QE/g).

3 Results

The analysis of the phenolic compounds composition in *B. alopecuroides* leaves using HPLC identify five components - vitexin (5) (t_R=12,0), isovitexin (6) (t_R=16,2), hyperoside (7) (t_R=18,3), isoquercitrin (8) (t_R = 19.8), rutin (9) (t_R = 20.6). An absorption spectrum of unidentified components has two bands, one of which (II band) is in the low-wave part (250-290 nm), the other - in the long-wave one (340-380 nm) (band I), that is a characteristic feature of the flavonoid structure [5]. All components were attributed to phenolic compounds based on these data. Table 2 provides data on the content of individual components in leaves of *B. alopecuroides* plants. The sample number corresponds to Table

Table 2. Phenolic compounds composition and content of *B. alopecuroides* plant leaves

Compounds		Sample No.									
No.	t_R	1	2	3	4	5	6	7	8	9	10
1	7.4	0	0.34	0	0	0	0	0	0	0	0
2	8.3	0	3.12	0.08	0.72	0	0	0	0.70	0	0
3	9.2	0.19	0.11	1.83	1.81	10.46	6.15	5.63	5.12	9.37	4.38
4	11.1	0.29	0	1.39	0.44	4.40	0.45	1.82	2.59	0	0.42
5	12.0	0.24	0	0	0.47	0.66	0	0	0.09	0	0
6	16.2	0.68	2.04	1.06	3.03	0.59	0.57	0.94	0.95	0.89	0.81
7	18.3	0.67	0	1.71	0	0.63	0	1.58	0.60	1.33	2.23
8	19.8	0.63	0	0.94	0	1.80	4.75	0.33	1.44	2.44	3.11
9	20.6	2.56	1.50	0.76	0.89	2.28	1.67	0.92	0.61	2.82	1.62
10	22.8	0	0	0	0	0	0	0	0	0	2.03
11	24.5	0	0	0	0	0	0	0	0	0	0.20
12	35.7	0	0.29	0.09	0	0	0.18	0.16	0	0.19	0.84
13	40.3	0	0	0	0.09	0	0	0	0	0	0
14	44.5	0	0.07	0	0	0	0	0	0	0	0
Σ		5.26	7.47	7.86	7.45	20.82	13.77	11.38	12.10	17.04	15.64

The relative standard deviation of repeatability in determining phenolic compounds was 0.011; t_R - retention time, min; No. of compounds: 5 - vitexin (t_R=12,0), 6 - isovitexin (t_R=16,2), 7 - hyperoside (t_R=18,3), 8 - isoquercitrin (t_R = 19.8), 9 - rutin (t_R = 20.6);

4 Discussion

B. alopecuroides as a typical mesophyte lives in humid, more or less solonetz meadows, swamp margins, stony slopes and pebbles, meadow steppes. It is spread mainly in Transbaikalia, except for highlands, eastward it is noted in the Amur River valley and Pre-Khanka lowland [6]. It was marked 14 components in total for *B. alopecuroides*, 5 of them - 1, 10, 11, 13, 14 - were sporadic (Table 2). The maximum composition (9 components) was noted in sample No. 10 collected in the Republic of Buryatia at 1140 m above sea level, the minimum composition (6 components) - in the leaves of plants No. 6 and 9 from the Blagoveschensk District. Routine, isovitoxin and unidentified component No. 3 (t_R = 9.2 min; UV spectrum - 273,355 nm) were found in all studied samples; hyperoside, isoquercitrine, astragaline and unidentified compound No. 4 (t_R=11,1) were frequently occurring compounds as well. The phenolic components composition in leaves of *B. alopecuroides* from different populations is quite variable, which may be due to both environmental factors (soil composition, microclimate, altitude, technogenic pollution), and to a possible hybridization of this species with the widespread in Eurasia *B. officinalis* Delarbre. This hypothesis on the hybridization of *B. alopecuroides* and *B. officinalis* Delarbre, and the origin of the species *B. elliptica* (Willd. Ex Spreng.) Kom as a result of such hybridization belonged to N. N. Tsvelev [6]. In general, the *Bistorta* genus species are characterized with a great polymosphism, they are often represented by numerous weakly differentiated races [7]. M.G. Popov [8] noted that *B. alopecuroides* in a typical form occurs only in "Proper Dauria: for example, Nerchinsk Plant and along the River Argun". Phenolic compounds are involved in many important processes of plant metabolism, and morphological variability is closely related to changes in chemical composition.

Variability was marked for the quantitative content of phenolic compounds as well. The smallest content of phenolic compounds (5.26 mg QE/g) was fixed in plant leaves collected in the Republic of Buryatia (No. 1) growing under the mixed forest canopy, the greatest one (20.8 mg QE/g) in plants from Tunka District in Buryatia (No. 5), which grew at 919 m above sea level. Plants gathered at 1140 m above sea level (No. 10) have a high content (15.64 mg QE/g) as well. Samples growing in Blagoveshchensk city surroundings (No. 6, 9) collected near the road contain up to 13.87-17.04 mg QE/g of phenolic compounds with the lowest composition of 6 components that is confirmed by numerous references on the enhanced synthesis of phenolic compounds, including flavonoids, under effect of various stresses. The flavonoid accumulation by plants is a mechanism to protect their photosynthetic apparatus from extensive oxidation under the influence of higher insolation or man-made contamination [9,10]. In the remaining samples the phenolic compounds content in leaves was about 7.5 mg QE/g.

In conclusion, the composition and content of phenolic components of *B. alopecuroides* from different populations of Siberia and the Far East is quite variable and depends on environmental factors, such as soil composition, microclimate, altitude, technogenic pollution. Three components were identified, which were found in *B. alopecuroides* plant leaves of all studied populations: isovitixine, rutin and unidentified component No. 3 (t_R = 9.2 min; UV spectrum 273,355 nm). *B. alopecuroides* plants growing in highlands and under conditions of technogenic pollution have the highest content of phenolic compounds that confirms the adaptive role of phenolic compounds in plant metabolism.

References

1. M. Adiba, I. H. M. Tabarak, Glob J. Res Med Plants Indigenous Med **2**, 669-674 (2013)
2. M. S. Voronkova, G. I.Vysochina, Chem. Sustainable Development **22**, 209-215 (2014).
3. E. P. Khramova, E. K. Komarevtseva, Vegetative Resources **44**, 96-102 (2008) In Russ.
4. T. A. Van Beek, J. Chromatography A **967**, 21-35 (2002)
5. E. P. Khramova, Chem. Vegetative Raw Materials **1**, 185-193 (2014.) In Russ.
6. N. N. Tsvelyov. *Vascular plants of the Soviet Far East* (4, Nauka, Leningrad, 1989) In Russ.
7. V. V. Petrovsky, *Arctic flora of the USSR* **5** (Nauka, Moscow, 1966.) In Russ.
8. M. G. Popov, *Flora of Middle Siberia* **2** (Nauka, Moscow, Leningrad, 1959) In Russ.
9. E. Grotewold, The science of flavonoids (Springer, New York, 2006)
10. M. S. Voronkova, G. I.Vysochina, Contemp. Probl. Ecology **6**, 565-574 (2013)

Analyses of asiatic hybrids resistance at introduction in northern forest steppe of west Siberia

Oksana Vronskaya[1,*], and *Natalia* Mosiahina[1]

[1]Federal Research Center for Coal and Coal Chemistry, Siberian Branch of the Russian Academy of Sciences, Kuzbass Botanical Garden, Russia

Abstract. The contents of main pigments of photosynthetic apparatus (chlorophyll a, b and carotinoid) in the leaves of Asiatic hybrids varieties during different stage of vegetation was first determined In conditions of northern forest steppe. It was established that 19 varieties of Asiatic hybrids had maximum level of accumulation of chlorophyll pigments (a and b) during the budding period, 4 – during the flowering period, 2 – during the fructification period; the most important value a/b was proper to 8 varieties in the budding period, 6 – in the fructification period and 11 in the flowering period, the maximum accumulation of carotinoind contained in the budding period that was expressed in the adaptive reaction of plants aimed to increase of photosynthetic apparatus resistance during the most dry season period. Analyses of photosynthetic pigment contents in the leaves of Asiatic hybrids varieties showed that the flowing changes in the contents of pigment complex may be considered as indicators of a mechanism of adaptation of lily varieties to the complex of climatic and natural conditions of northern forest steppe of West Siberia.

1 Introduction

The main task of botanical gardens activity is a search of new plants for its introduction in the culture. *Lilium* L. genus from *Liliaceae* Juss. family counts about 105 species and more then 5 thousand varieties. Asiatic hybrids come in the first group of International garden classification of lilies [1, 2]. They are known as decorative plant which used to create new varieties of lilies. This culture is under study in many regions of Russia and abroad [3–8].

Exchange of varieties of Asiatic hybrids between the botanical gardens allows to wide the range of varieties of collections, and to renew genofond. In such a case it's important that the plants adapt well for the complex of natural and climatic conditions of an introduction region. Much value in study of the plants condition has a studying of photosynthetic apparatus flexibility [9, 10] its adaptation to the varied conditions of introduction. The parameters of accumulation of photosynthetic pigments (chlorophyll a, b, carotinoind) contained in the leaves of Asiatic hybrids during different phase of vegetation

*Corresponding author: oksana_vronski@mail.ru

serve as the plants resistance indicators to the complex of natural and climatic conditions. And such parameter analysis gives a definite image on an adaptation of plants to the growth conditions. Firstly the contents of main pigments of photosynthetic apparatus (chlorophyll a, b and carotinoind) contained in the leaves of Asiatic hybrids varieties in different periods of vegetation in the conditions of northern forest steppe of West Siberia was defined and its indicator value. Studies objective: to define the contents of main pigments of photosynthetic apparatus (chlorophyll a, b and carotinoid) contained in the leaves of Asiatic hybrids varieties during different phases of vegetation, to define the varieties resistant tot the complex of natural and climatic conditions on the northern forest steppe of West Siberia.

2 Materials and methods

Studies were spent in 2012–14 years on the territory of Kuzbass botanical garden (Kemerovo city) located in the northern part of forest steppe area of West Siberia. Climate proper to the region under research – is sharply continental. The average monthly air temperature is about –0.9 °C. The higher temperature in summer is about +35 +38 °C, and the lower temperature in winter is – 57 °C. The first spring frosts begin in the period from August 26 till September 14 Average annual precipitation is about 450–500 mm, snow-cover height is from 47 up to 72 cm. Introduction populations – 25 varieties of Asiatic hybrids: 'Aelia', 'Virenea', 'Vostochnaya skazka', 'Zhizel', 'Zvezdochka', 'Kamilla', 'Kozette', 'Lastochka', 'Malinka', 'Narodnaya', 'Odetta', 'Poliushko', 'Polymia', 'Retro', 'Rozovaya dymka', 'Riabinka', 'Fleita', 'Yulia', 'Azurra', 'Alaska', 'Gold Lode', 'Nove Cento', 'Pepper', 'For you', 'White twinkle'. Number of a,b chlorophyll and sum of carotinoind was defined by the spectrophotometric method using the guidance papers [11, 12]. An analysis was made on the scanning device SPEK SSP–705. Statistical processing of finding was made using the PAST 3 device. Average error (±m) and coefficient of variation (V) were calculated per each parameter (M).

3 Results

Normal chlorophyll a, b and carotinoind ratio is: chlorophyll a – 50%, chlorophyll b – 30%, and carotinoind – 20% [13].

As L. pilosiusculum is a local species well adapted to the conditions of northern forest steppe of West Siberia, the figures of the pigments content in its photosynthetic apparatus were compared with the introduced varieties of Asiatic hybrids of out of region origin.

It was established that 19 varieties of Asiatic Hybrids had maximum accumulation of chlorophyll pigments (a and b) during the budding period, 4 – during the flowering period, and 2 – during the fructification period. The most contents of chlorophyll contained in the leaves of L. pilosiusculum (a, 1,178 ± 0,3 mg/g, b, 0,489 ± 0,03 mg/g). By compare the figures of chlorophyll contents of local species with the introduced it was determined that the contents of chlorophyll in the leaves of 'Lastochka', 'Pepper', 'For you' is near to L. pilosiusculum (a, 0,932–1,563 mg/g, b, 0,318–0,494 mg/g). The contents of chlorophyll in the leaves of 'Rozovaya dymka' (a, 0,501 mg/g, b, 0,203 mg/g) is in 2 times lower than the local species one. Intraspecific and intravarietal variability of chlorophyll contents is average (V = 0,1–9 %).

Quantity of chlorophyll a increased in 2,5–3 times the contents of chlorophyll b in all the phenophases of development during the lily vegetation period. The varieties coefficient of variation for the years of study amounted to V = 0,01–8,37 %. The highest index (V = 8,37

%) was proper to the variety 'Rozovaya dymka'. 'Aelita' variety showed a stable contents of chlorophyll a and b V = 0,01 %.

One on informative figures characterizing the photosynthetic apparatus work is a correlation of chlorophyll a to chlorophyll b (a/b). Such correlation is caused by the «main» chlorophyll a activity. The higher activity increases more the photosynthetic rate. Normally such a figure correspond to 2,2–3,0 mg/g of green weight [14].

Following the studies it was determined that 8 varieties had the highest value a/b in the budding period, 6 – in the fructification period and 11 in the flowering period. By compare the average figures a/b it was established that the varieties 'Lastochka' had the figures (3,281 mg/g), and 'Polymia' (3,020 mg/g) which are higher than such the figures proper to the variety L. pilosiusculum (2,319 mg/g og green weight). Quantity a/b of all the varieties under study was within the standard values certifying the normal photosynthetic rate. The highest coefficient of a/b variation in the flowering period was proper to the variety 'Fleita' V = 4,01 %, the lowest V=0,1 % was proper to the variety 'Aelita'.

It's known that carotinoid is an important element of lily's pigment system. It was determined that the most part of the introduced varieties of Asiatic hybrids showed the maximum accumulation of carotinoid in the budding period, expressed in the adaptive reaction of the plants aimed to increase resistance of photosynthetic apparatus in more dry season period. Quantitative content of carotinoid in the leaves of L. pilosiusculum (0,368 mg/g) was significant equal to the varieties 'Lastochka' (0,305 mg/g) and 'Kozette' (0,393 mg/g) and higher in 1,7 times than contained in the leaves of 'Vostochnaya skazka' (0,221mg/g).

Correlation a+b/carotinoid is no less important in characteristic of photosynthetic apparatus work. This correlation is stable in the norm and is sensitive to any changes in different environment factors. As L. pilosiusculum as the introduced varieties showed a tendency to reduce the correlation a+b/carotinoid near to fructification period that reveal reducing of light-harvesting function of pigment complex under unfavorable temperature conditions impact. At average we got high values a+b/carotinoid demonstrating the good adaptation mechanism of Asiatic hybrids to the environment.

Correlation relationship reflects well the plants state as within the usual conditions as within the stresses. Studying intercorrelation of the photosynthetic pigments content we determined that all the varieties have the similar correlation dendrites and the Pleiades order. Total number of authentic correlation relationship ($r \geq 0,601$) is variable from 10 to 20. The high level of the parameters conjugation took place in 2013. The close intercorrelation between the studied signs is discernible: carotinoid, chlorophyll a+b, chlorophyll a/b; chlorophyll a and chlorophyll b (fig. 1).

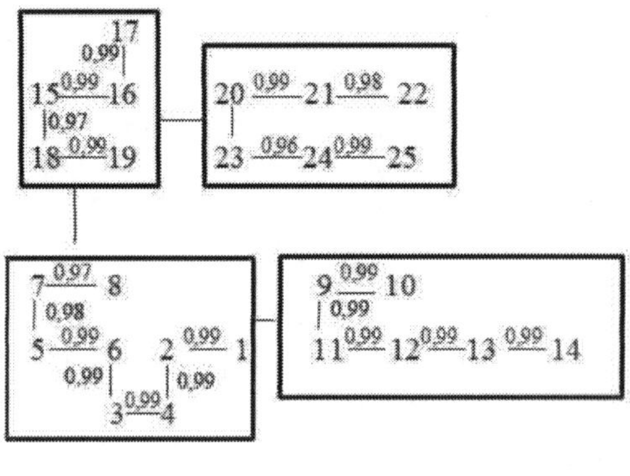

Fig.1. Correlation dendrite content of the main pigments of the photosynthetic apparatus (chlorophylls a and b, carotenoids) in the leaves of varieties of Asian hybrids. 1. 'Aelita', 2. 'Virineya',

3. 'Vostochnaya skazka', 4. 'Zhizel', 5. 'Zvezdochka', 6. 'Kamilla', 7. 'Kozette', 8. 'Lastochka', 9. 'Malinka', 10. 'Narodnaya', 11. 'Odetta', 12. 'Poliushko', 13. 'Polymia', 14. 'Retro', 15. 'Rozovaya dymka', 16. 'Riabinka', 17. 'Fleita', 18. 'Yulia', 19. 'Azurra', 20. 'Alaska', 21. 'Gold Lode', 22. 'Nove Cento', 23. 'Pepper', 24. 'For you', 25. 'White twinkle'

On the dendrogram of varieties similarity over the pigment content built using the cluster analyses, the species family division is shown. Asiatic hybrids form a single cluster with a shoot height up to 75 cm, another cluster includes the lilies with a culm length up to 75 cm, and the other cluster includes the lilies with a culm length of 80 cm and more (fig. 2).

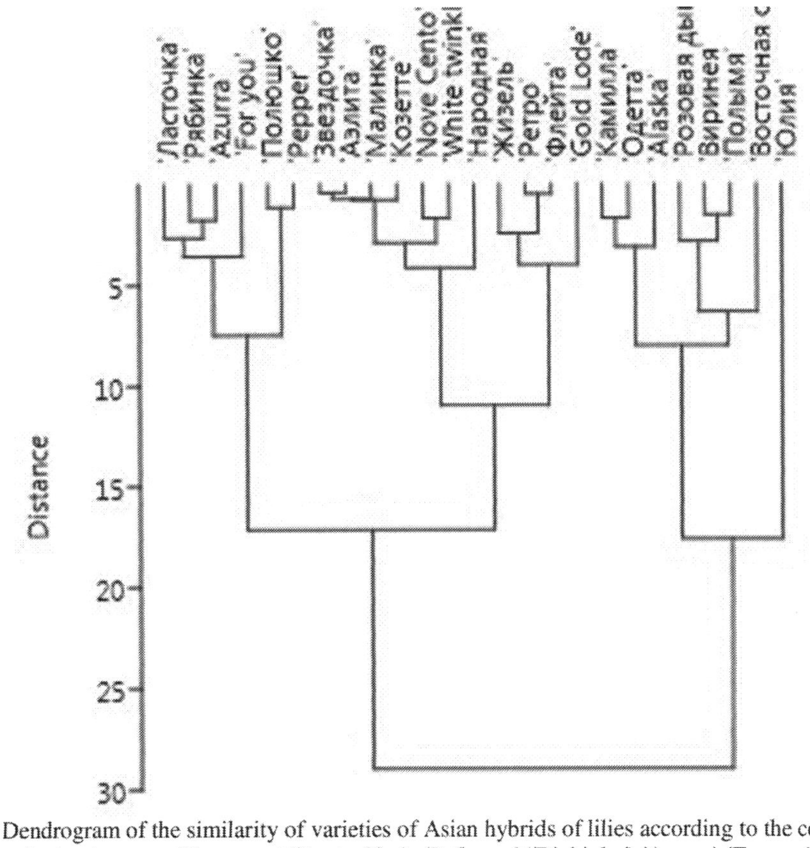

Fig. 2. Dendrogram of the similarity of varieties of Asian hybrids of lilies according to the content of photosynthetic pigments. 'Ласточка' ('Lastochka'), 'Рябинка' ('Riabinka'),'Azurra', 'For you', 'Полюшко' ('Poliushko'), 'Pepper', 'Звездочка' ('Zvezdochka'), 'Aelita', 'Малинка' ('Malinka'), 'Козетте' ('Kozette'),'Nove Cento', 'White twinkle', 'Народная' ('Narodnaya'), 'Жизель' ('Zhizel'), 'Ретро' ('Retro'), 'Флейта' ('Fleita'),'Gold Lode', 'Камилла' ('Kamilla'), 'Одетта' ('Odetta'), 'Alaska', 'Розовая дымка' ('Rozovaya dymka'), 'Виринея' ('Virineya'), 'Полымя' ('Polymia'), 'Восточная сказка' ('Vostochnaya skazka'), 'Юлия' ('Yulia').

4 Discussion

Thus, the content of photosynthetic pigments contained in the leaves of Asiatic hybrids showed that the passing changes in the pigment complex structure may be considered as indicators of the mechanism of adaptation of lilies varieties to the complex of natural and climatic conditions on the northern forest steppe of West Siberia. The most varieties meet norms by every measure. It means that the formedness of photosynthetic apparatus of plants and its work ensured availability of a strong protective pigment system that allowed to the

Asiatic hybrids varieties to develop well within the vegetation season. Analyses of contemporary foreign literary sources shows that the photosynthetic activity of the eastern hybrids of lilies increases in shade and reduces in water deficit [15, 16], i.e. is in complete accord with our opinion concerning protective function of photosynthetic apparatus. Based on the research findings we selected the varieties of Asiatic hybrids with well developed pigment system: lilies of subdivision I. Asiatic hybrids: 'Aelita', 'Virenea', 'Zvezdochka', 'Kamilla', 'Odetta', 'Poliushko', 'Retro', 'Yulia', 'Alaska', 'Gold Lode', 'Pepper', '4 you'.

References

1. The Internat. Lily Register, 377 (1982).
2. The Internat. Lily Register end checklist The Royal Horticultural Society, 948 (2007).
3. L.V. Zavadskaya. Modern lines of activity of botan. Gard. and botan. Collect., on the conservation of plant biodiversity, 241 – 244 (2005).
4. L.M. Kikot. Modern trends in the activities of botanical gardens and botanical collections for the conservation of plant biodiversity: Sat. materials, 99 – 100 (2005).
5. K.B. Lim, M. Jaap van Tuyl. Lily. Lilium hybrids, 300 (2006).
6. S. Guven, S Okur, M Demirel, K. Coskuncelebi, S. Makbul1, O. Beyazoglu. Biol, **69/9**, 1122–1133 (2014).
7. K. Lim., A. Younis, J. Taek Park, Y. Jung Hwang. Not Sci Biol. **6/2**, 178–184 (2014).
8. O.O. Vronskaya. [Electronic resource]. Modern problems of science and education. 6 2015. http://www.science-education.ru/ru/article/view?id=23469.
9. M.M. Chaves, J. Flexas, C. Pinheiro. Ann. Bot, **103**, 551–560 (2009).
10. K. Inamoto, K. Nagasuga, T. Yamazaki Yano, J. Hort. Sci. Biotech. **90**, 259–266 (2015).
11. A.A. Shlik. Modern problems of photosynthesis **85**, 108 (1973).
12. H.K. Lichtenhalter, A.R. Wellburn, Biochem. Soc. Rans. **11**, 591–598 (1983).
13. E.V. Maksimova A.A. Kosicina O.N. Makurina, Vestnik Sam GU. **8**, 146–152 (2007).
14. M.S. Titova, Vestnik OGUK. **12**, 9–12 (2010).
15. Y.J. Zhang, Z.K. Xie, Y.J. Wang, P.X. Su, L.P. An, H.Gao, Rus. J. Plant Physiol. **58**, 844 (2011).
16. Y.J. Zhang, Z.K. Xie, Y.J. Wang, P.X. Su, L.P. An, H.Gao, Rus. J. Plant Physiol. **62**, 334–339 (2015).

Effect of prolonged cultivation on the micropropagation *in vitro* of mint cultivars and breeding samples

Margarita Zagorskaya*, and *Natalia* Yegorova

FSBSI "Research Institute of Agriculture of Crimea", Simferopol, 295493, Russia

Abstract. The influence of the genotype and the number of subcultures on the development of meristem cultures of five mint cultivars and breeding samples – Krasnodarskaya 2, Diana, K37, No. 2.8.14, A_2 (*Mentha* spp.) was studied. At the second stage of micropropagation, the analysis of morphometric parameters of explants during six subcultures of four genotypes showed an increase in the multiplication index at 4-5th passages, while in the sample A_2 - at 3-4th passages.

1 Introduction

Mint is a well-known perennial herbaceous medicinal, essential oil and spicy aromatic plant in the family Lamiaceae. Different species of the genus *Mentha* are used in pharmacology, cosmetic and food industries, as well as in medicine. Preparations from mint plants have antiseptic, sedative, spasmolytic, and choleretic effect. Great interest to this valuable plant causes active development of breeding work, including creation of new essential oil cultivars. It is reasonable to involve biotechnological methods for the rapid propagation of the initial breeding material and cultivars, as well as for the development of techniques for their preservation *in vitro*.

The literature sources contain data regarding the studies related to the development of micropropagation methods for various mint species - *Mentha piperita* [1-3], *M. viridis* [4], *M. canadensis, M. suaveolens, M. longiflora, M. aquatica, M. arvensis, M. x gracilis, M. gracilis, M. spicata* [5]. According to these sources, scientists used meristems, buds [1, 3, 4] and stem segments with a node [2, 5] as explants in their experiments, as well as used different modifications of culture medium composition at the stages of propagation. However, many important issues of clonal micropropagation were not risen in these works, especially for cultivars and breeding samples of essential oil mint grown in the Crimea. The aim of our research was to study the influence of the genotype and the number of subcultures on the development of mint explants at the second stage of micropropagation *in vitro*.

2 Materials and Methods

* Corresponding author: zagorskayamargo@gmail.com

Tissues and organs of mint cultivars and breeding samples – cultivar Diana (interline hybrid *Mentha cadadensis* L.), cultivar Krasnodarskaya 2 (obtained by selecting in the seed progeny of allopolyploid sample *Mentha x piperita* L.), sample K 37 (*M. rotundifolia* L. Huds), sample No. 2.8.14 (line S_1 of *M. spicata* K65 sample) and sample A_2 (allopolyploid sample *M. x piperita* L.) [6] were used as the material for the study. Meristems with one pair of leaf primordia isolated from axillary buds were introduced in culture *in vitro*. At the 2nd stage of micropropagation, segments of the stem with a node isolated from shoots, which were developed from meristems, were used as explants. Explants were cultivated on Murashige and Skoog culture medium (MS) supplemented with growth regulator such as BAP, Kinetin, IAA, GA_3 (Sigma, USA). Explants in test tubes were kept at 26°C, 70% of relative air humidity, under 16-h photoperiod with illumination 2-3 klx.

Microcuttings were conducted every 40-45 days, before which the length and number of shoots, the number of nodes (pairs of leaves) on shoot, the frequency of rhizogenesis were determined. The multiplication index was calculated as the number of microcuttings that can be obtained during one subculture. For this the number of formed on the explant shoots was multiplied on the number of nodes on the shoot. Experiments on *in vitro* culture were carried out in three replications; at each variant 20 explants were analyzed. The data are processed using the Microsoft Office software package (Excel 2007). The tables show average values and their standard errors.

3 Results and Discussion

The studies revealed that active adventitious shoot formation and development of shoots which are long enough (up to 20-40 mm) were observed for almost all genotypes (excluding A_2) when introducing meristems in culture *in vitro*. For the further micropropagation at the 2nd stage, microcutting of shoots was carried out. Microcuttings (segments of a stem with one node and a pair of leaves) were cultivated on the modification of MS culture medium with 1.0 mg/l mg L^{-1} BAP and 0.5 mg L^{-1} IAA, which we used for the propagation of other cultivars of peppermint [7]. It was identified that this medium was effective for four genotypes, whereas explants of sample A_2 did not develop on it. Therefore, modification of MS medium with 0.5 mg L^{-1} Kinetin and 0.5 mg L^{-1} GA_3 was chosen for this sample in a separate experiment.

During microcuttings cultivation, on average developed 2.7-15.2 shoots per explant, depending on the number of subcultures and genotype (Table 1.). At the same time, active growth of axillary and adventitious shoots was observed. They reached a length of 15.2-49.7 mm after 40-45 days of cultivation. Rhizogenesis with a frequency of 7.1 to 100% was observed for all studied genotypes at the second stage of propagation, during which an average of 1.0-5.7 roots per explants were formed. It was revealed that it is expedient to combine two methods for mint micropropagation *in vitro* – the induction of adventitious shoots and microcutting of main and additional shoots.

When analyzing the development of explants during six passages, the peculiarities of the influence of genotype and number of subcultures on the morphometric parameters of microshoots and multiplication index were identified (Table 1.). Gradual increase in the number and length of shoots and in the number of nodes during the $2\text{-}5^{th}$ passages was observed for cultivars Krasnodarskaya 2 and Diana. Rhizogenesis with a frequency of 95.0 to 100% was observed for these passages. The main parameter at the second stage of micropropagation – multiplication index was significantly increased; it reached a maximum (23.0-48.7) for these cultivars in $4\text{-}5^{th}$ passages. The decline of almost all the studied parameters occurred then in the 6th passage.

Table 1. Influence of subculture number and genotype on the development of mint explants on the second stage of micropropagation *in vitro*

Number of subcultures	Cultivar, sample	Shoots number, pcs.	Shoots length, mm	Nodes number, pcs.	Rooting frequency, %	Multiplication index
1	Krasnodarskaya 2	3.1±0.6	15.2±1.2	1.9±0.1	7.1±0,7	6.1±1,1
	No. 2.8.14	5.9±0.6	41.0±3.0	2.9±0.2	64.2±5,5	17.6±2,7
	Diana	4.3±0.7	33.0±2,2	2.7±0.1	31.2±2.2	11.8±2.0
	K 37	4.3±0.4	39.9±3.3	4.3±0.3	93.7±8.8	19.1±1.5
	A_2	3.4±1.3	33.5±3.2	3.1±0.2	42.8±5.0	10.7±1.9
2	Krasnodarskaya 2	4.6±0.7	27.1±2.1	3.0±0.2	100.0	14.2±2.3
	No. 2.8.14	4.0±0.3	44.7±3.0	3.6±0.2	91.3±9.0	14.9±1.0
	Diana	4.9±0.3	44.9±2.7	3.4±0.1	95.8±8.9	17.0±1.3
	K 37	6.0±0.4	26.7±1.8	2.8±0.1	100.0	17.1±1.1
	A_2	2.7±0.4	41.5±4.8	3.5±0.3	90.0±8.8	9.7±1.0
3	Krasnodarskaya 2	12.3±1.0	25.1±1.1	2.2±0.1	60.8±5.9	27.7±2.7
	No. 2.8.14	4.3±0.4	33.7±3.1	3.2±0.2	91.3±8.9	14.3±1.0
	Diana	5.3±0.5	36.5±2.4	3.3±0.1	100.0	17.7±1.7
	K 37	5.3±0.3	38.3±2.5	3.9±0.2	100.0	21.3±1.9
	A_2	4.3±0.4	39.1±3.6	2.7±0.2	90.0±8.7	12.0±1.0
4	Krasnodarskaya 2	15.2±0.7	34.1±1.1	3.2±0.1	100.0	48.7±3.2
	No. 2.8.14	3.6±0.6	42.0±3.9	3.7±0.3	56.5±5.3	13.8±1.4
	Diana	6.5±0.7	38.9±2.3	3.8±0.2	100.0	24.7±2.1
	K 37	5.3±0.3	34.9±2.5	3.7±0.2	100.0	19.9±1.7
	A_2	3.4±0.3	44.0±4.1	3.5±0.3	80.0±7.6	12.1±2.0
5	Krasnodarskaya 2	10.0±0.8	49.7±2.1	4.7±0.2	95.2±8.5	47.3±4.9
	No. 2.8.14	4.8±0.6	48.7±3.5	4.1±0.2	95.2±8.9	20.1±2.7
	Diana	6.2±0.4	46.0±2.6	3.7±0.1	95.2±8.9	23.0±2.1
	K 37	6.4±0.4	40.1±2.2	3.6±0.1	100.0	23.3±2.2
	A_2	2.2±0.2	41.7±4.4	2.8±0.1	60.0±5.4	6.2±0.6
6	Krasnodarskaya 2	4.5±0.7	30.0±2.5	2.4±0.1	5.0±0.2	10.9±1.6
	No. 2.8.14	3.8±0.4	48.2±3.8	3.6±0.2	60.0±5.8	14.0±1.5
	Diana	3.9±0.4	42.1±3.1	2.6±0.1	45.0±3.9	10.3±1.1
	K 37	4.8±0.5	39.4±3.1	3.5±0.2	65.0±6.6	17.2±1.6
	A_2	2.1±0.2	46.4±4.9	3.9±0.4	75.0±6.8	8.4±1.1

Similar tendencies of development were observed in other mint genotypes, however, there were no significant differences in many parameters during the 2-5th passages. However, the maximum multiplication index was identified for samples No. 2.8.14 (20.1) and K37 (23.3) in the fifth passage. After that, a slight decrease in the morphometric parameters was observed in the sixth passage. The maximum number of shoots, rooting frequency and multiplication index (12.0-12.1) were noted in the 3-4th subcultures for sample A_2. Then, there was a significant decrease in parameters, and in the fifth passage the multiplication index was only 6.2 for this genotype.

Gomes et al. [5] when studying nine mint species during five passages also showed an increase in the multiplication index after the first passage. It reached the highest values in 3-5th subcultures, depending on the species. The same situation of increasing the multiplication index to a certain passage during prolonged micropropagation was revealed for some other essential oil plants. Thus, at the 2nd stage of micropropagation *in vitro* an increase in the multiplication index, which reached maximum in the third passage, was observed for the cultivars of essential oil rose [8] and lavender [9]. At the same time, this parameter for geranium was stable for two years, but for sage and fennel did not change significantly during the first three subcultures, and then it gradually decreased [9].

Comparing different mint genotypes, it was shown that 'Krasnodarskaya 2' had the best micropropagation ability. This cultivar showed the most active adventitious shoot formation, and as a result the number of shoots reached maximum value (15.2 per explant) in the 4th passage. The largest average length of shoots and the number of nodes were in the 5th passage. As a result, multiplication index for this cultivar reached the highest value – 48.7. The parameters of explants development for sample A_2 were the lowest. Its multiplication index varied from 6.2 to 12.1, depending on the passage.

Thus conducting mint propagation during six subcultures, it was established that the multiplication index was increased in the 4-5th passages for cultivars Krasnodarskaya 2, Diana and samples K37 and No. 2.8.14. But for micropropagation sample A_2 it was neccesary to use culture medium of another composition, on which this index reached a maximum in 3-4th passages.

References

1. G.J. Minas, Acta Hort. **853**, 77-82 (2010)
2. J. Mehta, R. Naruka, M. Sain, A. Dwivedi, D. Sharma, J. Mirza, Asian J. Plant Sci. Res. **2** , 518-523 (2012)
3. I.A. Bugara, Scientific Notes of Taurida V.I. Vernadsky National University, Ser. Biology, chemistry **26**, 10-15 (2013)
4. H.D. Raja, D.I. Arockiasamy, Plant Tissue Cult. & Biotech. **18**, 1-6 (2008)
5. H.T. Gomes, P.M. Bartos, A.E. Marins, S.O.D. de Oliveira, J.E. Scherwinski-Pereira, Acta Scientiarum. Biol. Sci. **37**, 405-410 (2015)
6. L.A. Bugaenko, N.P. Shilo, *Polyploidy and interspecies hybridization in mint* (Simferopol, Business-Inform, 2012)
7. M.S. Zagorskaya, N.A. Yegorova, *Proceed. Int. Conf. dedicated to 85th anniversary of the Central Bot. Gar. of the Nat. Academy of Sci. of Belarus* **2**, 205-208 (Medisont, Minsk, 2017)
8. N.A. Yegorova, I.V. Stavtzeva, Bull. of the State Nikit. Botan. Gard. **120**, 36-43 (2016)
9. N.A. Yegorova, A.G. Krivokhatko, I.V. Stavtzeva, L.I. Kamenek, Taurida Herald of the Agrarian Sciences **1,** 9-14 (2013)

Physiological and biochemical characteristics of *Populus tremula* leaves in anthropogenic disturbed habitats

Yulia V. Zagurskaya[1,*], *Tatyana I.* Siromlya[2], *Olga V.* Kotsupiy[3], *Elena Yu.* Babaeva[4,5], and *Vladimir I.* Ufimtsev[1]

[1] Federal Research Center on Coal and Coal Chemistry SB RAS (Institute of Human Ecology), 650065, Kemerovo, Russia
[2]Institute of Soil Science and Agrochemistry SB RAS, 630090, Novosibirsk, Russia
[3]Central Siberian Botanical Garden, SB RAS, 630090, Novosibirsk, Russia
[4]People's Friendship University of Russia, 117198, Moscow, Russia
[5]All-Russian Sci. Research Institute of Medicinal and Aromatic Plants, 117216, Moscow, Russia

Abstract. The main morphological, anatomical, physiological and biochemical characteristics of *Populus tremula* undergrowth leaves as resistance indicators under conditions of coal pit waste dump are considered. There has been established high variability of morphological parameters and chemical elements composition and anatomical structure peculiarities. There are reported the differences in the content of photosynthetic pigments and phenol compounds.

1 Introduction

The study of plant resistance and adaptation mechanisms to anthropogenic disturbed habitat conditions is one of the most urgent directions of modern botanic and ecology. The most important objects under study among woody plants are the main forest forming kinds. In Siberia, they are hardwood of *Betula* and *Pópulus* including *Pópulus trémula* [1]. Resistance mechanisms of *P. tremula* to difficult environments and anthropogenic impact are studied insufficiently [2-8].

The goal was to study some biochemical and physiological characteristics of aspen leaves under industrial pollution conditions.

2 Materials and Methods

2.1 Object of study

The object of the study was *Pópulus trémula* coenopopulations of 10-15 years on the territory of "Kedrovskiy" coal pit (Kemerovo city) under ontogenesis generative stage

* Corresponding author: syjil.extra@gmail.com

(undergrowth). The samples of laminas were collected in the beginning of August 2015 from brachyblasts of different trees at 1.5-2 m height. The experiment plots were: 1) the territory with virgin soil layer without intensive transport stress (control); 2) the overgrowing dump of about 30 years (dump 1); 3) the aspen-reclamated dump of about 30 years with in 50 m from the dump bank of working pit (dump 2); 4) the technological road border on the coal pit territory with regular traffic of mining technical equipment and large-size diesel transport (road).

2.2 Physiological and Chemical methods

The total water content in the leaves and extractive substances amount were determined by gravimetric method [9]. Linear parameters were measured with a ruler (SS 17435). The leaf square was calculated by carving method. Anatomic leaf structure was studied by the methodology reported in State Pharmakopoeia of the RF. XIII ed. (OFS 1.5.1.0003. 15) [9], LOMO Mikmed-1 microscope with AU-12 binoculars of 1.5x and 1.8x (ocular 10x, objective lens 10x, 20x and 40x) and MBC 10 binocular as well.

Photosynthetic pigments content was determined by spectrophotometric method [10-12], phenol compounds composition and content - by HPLS method [13]. Chemical elements composition was defined after dry ashing by atomic-emission spectroscopy method and accuracy control on the standard samples. Variance analysis was carried out by Kruskal-Wallis method and correlation one - by Spearman method, critical level of p values was taken as 0.05. All data are given in terms of overdry matter.

3 Results

The leaves of aspen brachyblasts were round-shaped (up to 50%) or egg-shaped (about 20%), there were rarely occurred elliptic (rhomb)-shaped, inversely egg-shaped or asymmetric ones; there were not revealed statistically significant differences in lamina shapes. The length (M/Min-Max: 5.2/3.8-7.0 см), the width (4.9/3.5-6.5 см) and the leaves square (23.3/18.8-31.8 $см^2$) varied significantly but did not differ in depend on habitat either.

The studied samples had similar anatomical characteristics. The cells of abaxial and adaxial leaf epidermis had anfractuose walls. Stomatal apparatus is anomocytic type. Around stomata there was observed cuticle rugosity. Lamina edge was slightly wavy, nervation in mesophill was seen clearly and on the large veins there were occurred prismatic crystals of CaC_2O_4 (Fig. 1).

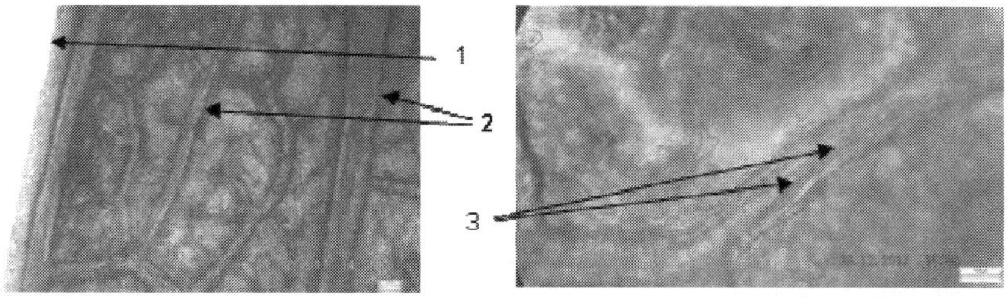

Fig. 1. Characteristics of anatomical structure of aspen leaf (1 – lamina edge (Magnification 450), 2 – veins, 3 – crystals of CaC_2O_4 on the veins (Magnification 900)).

The total water content in the leaves from different habitats was within 51-59% that was comparable with the values for the most mesomorphic plants. With anthropogenic stress growing, the extractive substances content, probably, increased due to hydrophilic fraction compounds (Table 1).

In aqueous-alcoholic extracts of aspen leaves we revealed from 20 to 27 compounds of phenol origin with 4 dominating flavonol glicosides (including hyperoside, rutin and astragaline) and 6 components of phenol carbonic acids (including gallic acid, clorogenic and cinnamic ones).

Table 1. Content of extractive substances, photosynthetic pigments and phenol compounds in aspen leaves (M±S), % per overdry weight.

Content	Control	Dump 1	Dump 2	Road
Total soluble matters (acetone)	11,8±0,5	11,7±0,4	12,2±0,3	15,3±0,8
Hydrophilic compounds	5,9±0,8	7,5±0,6	7,4±0,6	10,4±0,7
Chlorophylls	0,52±0,06	0,40±0,03	0,35±0,04	0,42±0,03
Carotenoids	0,17±0,02	0,14±0,01	0,12±0,01	0,15±0,01
Flavonoids	3,1±0,2	2,2±0,2	4,0±0,2	2,7±0,2
Phenolic acids	0,53±0,09	0,81±0,14	1,31±0,27	1,62±0,23

Chemical elements composition of *P. tremula* leaves from disturbed habitats did not differ. While comparing with the samples taken from the control plot, there were not revealed statistically significant difference in the content of the main plant nutrition elements of N, P, K. "Microelements" concentration (mg/kg) also changed within the same values: B (40-170), Co (0.9-1.6), Cu (5-8), Mo (0.1-0.4) and Zn (120-260) as well as that of some other chemical elements including "toxic" ones: Sn (0.4-1.2), Cd (0.7-1.9), Ni (2.5-4.8), Sr (40-170) and Mg (1800-4200) (Fig. 2).

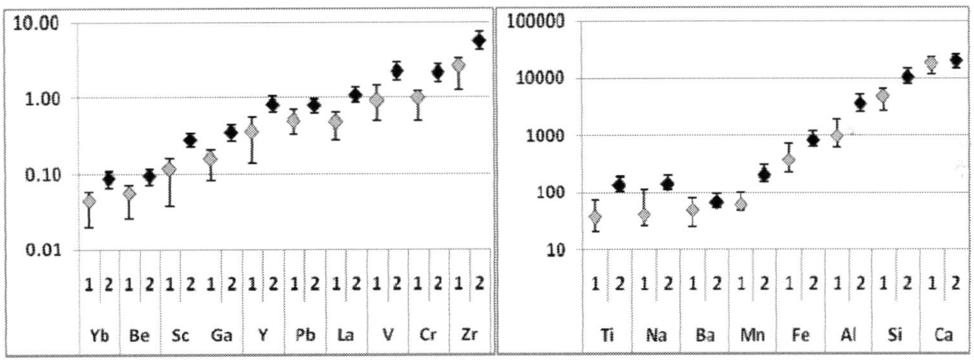

Fig. 2. Content of some chemical elements of *P. tremula* leaves from disturbed habitats, mg/kg (1 – disturbed ecotopes, 2 – control plot).

4 Discussion

The study results of anatomic and morphological characteristics of the aspen undergrowth brachyblasts leaves in the conditions of "Kedrovskiy" coal pit generally correlate with the reported data [14], but they are more variable. The total content of photosynthetic pigments in the aspen leaves from anthropogenically disturbed habitats decreased that confirmed reported information [4-6], besides that, there were observed increasing of pigments share in the amount of substances extracted with weakly polar extractant (acetone) [3].

According to reported data, in the aspen leaves there were revealed more than 20 of phenic compounds, including flavonoids (hyperoside, rutin, quercitrin, isoquercitrin, astragalin) and also phenolglicosides; hydroxibenzoic and hydroxicinnamic acids [7, 8 14-17]. The peculiarity of phenol compounds composition of anthropogenically disturbed ecotopes was ratio change of hyperoside and rutin dominating flavonoids.

There was revealed direct strong correlation of rS – 0.75-0.93 between all chemical elements content being of higher amounts in the plants of the control plot and studied leaves ashing. Increased dust level, probably, was connected with granulometric soil composition; the content of highly dispersed clay fraction reached 45-48% while on the other plots it did not exceed 3-10% and was accompanied by a very high rubbliness (30-40%) that was specific to immature soils of anthropogenic landscapes [18]. There also has been reported that technogenic pollution impact is not the only significant environmental factor that forms a certain level of biomass chemism as quantity changes depend more on the specific habitat conditions [19]. Chemical elements composition of *P. tremula* in comparison with reported data shows as similarities so differences [20].

References

1. I.Yu. Koropachinskiy, T.N. Vstovskaya, *Woody plants of the Asian part of Russia* (Novosibirsk: Publ. House of SB RAS, Branch «Geo», 2002)
2. V.V. Tuzhilkina, Rastit. Resursy **51**, 181 (2015)
3. Y.V. Zagurskaya, Reg. Environm. Iss. **1**, 24 (2017)
4. Yu.V. Zagurskaya, Sib. J. Forest Sci. **1**, 105 (2017)
5. M.V. Tcherkashina, G.A. Petukhova, Adv. Curr. Nat. Sci. **2**, 41 (2007)
6. A.A.Rostunov, T.A. Konchina, Bull. Irkutsk St. Univ. Ser.Biol., Ecol. **15**, 68 (2016)
7. K. Nissinen, V. Virjamo, T. Randriamanana, N. Sobuj, U. Sivadasan, L. Mehtatalo, E. Beuker, R. Julkunen-Tiitto, L. Nybakken, Canad. J. Forest Res. **47**, 1350 (2017)
8. N. Sobuj, V. Virjamo, Y. D. Zhang, L. Nybakken, R. Julkunen-Tiitto, Environm. & Exp. Bot. **146**, 34 (2018)
9. *State Pharmakopoeia of the RF. XIII ed.* (Moskow, Min. Zdrav. RF, 2015)
10. G. Holm, Acta. Agr. Scand. **4**, 457 (1954)
11. D. Wettstein, Exp. Cell Res. **12**, 427 (1957)
12. N.V. Shadrin, I.V. Drobetskaya, I.N. Chubchikova, N.V. Terentyeva, Mar. Ekol. J. **7**, 85 (2008)
13. O.V. Kotsupiy, G.I. Vysochina, Schol. Not. Transbaikal St. Univ. **11**, 104 (2016)
14. Y.V. Banayev, S.V. Shishkin, M.S. Voronkova, A.P. Belanova, M.A. Tomoshevich, Bull. Altai St. Agric. Univ. **8** , 90 (2017)
15. E. Häikiö, M. Makkonen, R. Julkunen-Tiitto, J. Sitte, V. Freiwald, T. Silfver, V. Pandey, E. Beuker, T. Holopainen, E. Oksanen, J. Chem. Ecol., **35**, 664 (2009)
16. I.Yu. Lobanova, J. Ural Med. Acad. Sci. **35**, 67 (2011)
17. I.N. Abreu, M. Ahnlund, T. Moritz, B.R. Albrectsen, J. Chem. Ecol. **37**, 857 (2011)
18. V.A. Androkhanov, E.D. Kulyapina, and V.M. Kurachev, *Soils of Technogenic Landscapes: Genesis and Evolution* (Izd. SO RAN, Novosibirsk, 2004)
19. L.M. Kavelenova, A. Zdetovetskii, A. Ognevenko, Khim. Rastit. Syrya, **3**, 85 (2001)
20. R.H. Giniyatullin, Agr. Bull. Urals, **82**, 80 (2011)

Isomer composition of tocopherols in seabuckthorn (*Hippophae rhamnoides* L.) seeds in forest-steppe area of Altai territory

Anna Zemtsova[1,*], and *Yury* Zubarev[1]

[1]Federal Altai Scientific Center of Biothechnology, Barnaul, 659910, Russia

Abstract. Estimation of tocopherol isomers composition in seed oil of seabuckthorn (*H. rhamnoides* ssp. *mongolica*) varieties is represented. It has been established that α-tocopherol is the dominant tocopherol in the lipids of the seabuckthorn seeds, the mean content of which for two years has been ranged from 2.1 (Dar Katuni) to 4.8 mg/100 g (Yantarnaya). In general within the evaluated group of subspecies, the mean content of α-tocopherol was determined at 3.5±0.3 mg/100 g. β-, γ- and δ-tocopherols were characterized by low content, accumulating an average for two years at 0.5±0.1, 1.3±0.1 and 0.7±0.3 mg/100 g, respectively.

1 Introduction

Tocopherols (vitamin E) considered as one of the most important substance of lipophilic origin. They possess vitamin activity, have significant antioxidant properties that prevent lipid peroxidation by inactivating free radicals of fatty acids [1]. It has been found out that tocopherols in plants are represented by closely related by chemical structure isomers – α-, β-, γ-, δ-tocopherols [2]. They are only producing by plants and convey to human body with food, mainly as vegetable oil [3].

Seabuckthorn (*Hippophae* L.) is one of the most prospective horticultural crop, which synthesize oil both in pulp and seeds and widely used in industry. Tocopherols are quite valuable part of seabuckthorn oil. Several investigations on tocopherol isomers content in seabuckthorn pulp oil have been already done on some subspecies. Meanwhile till now there have not been carried out any researches on tocopherol isomers estimation in seabuckthorn seeds of *H. rhamnoides* subspecies *mongolica*, which widely grown in Siberia and considered as most promising subspecies within genus *Hippophae* L. This subspecies is represented by many so called ecotypes, which are grown in different regions, soil and climate conditions. The reason of current investigation was comparative estimation of tocopherol isomers composition in seed oil of different ecotypes of *H. rhamnoides* subspecies *mongolica*, growing in the same soil and climate conditions on experimental plots of Lisavenko Research Institute of Horticulture for Siberia.

* Corresponding author: anna-krysova@mail.ru

2 Materials and Methods

Research has been carried out on facility of UBF GmbH laboratory (Germany) in 2014-2015. Samples have been collected on experimental plots of the Lisavenko Institute of Horticulture for Siberia in forest-steppe area of Altai Territory. All samples till evaluation kept for 6 months at temperature -25^0C. As objects of investigation the following varieties have been taken: Dar Katuni, Novost Altaya – Katun ecotype; Chuyskaya – Chuya ecotype; Inja, Elizaveta – mutants (obtained by chemical mutagenesis); Velikan, Yantarnaya – Sayan-Katun ecotype; Zhivko, 42-68-2 – Krasnoyarsk-Sayan ecotype; Lubimaya, Chulyshmanka – Sayan-Chulyshman ecotype; Zarja Dabat – Buryat ecotype.

Tocopherols estimation carried out by using Shimadzu Liquid Chromatograph LC-6 A, equipped by spectrophotometric detector Shimadzu SPD-6 AV. Tocopherols were identified by retention time and quantified using an external standard based on measurements of peak area values. As standards α-, β-, γ- and δ-tocopherols have been used (Bestimmung der Tocopherole und Tocotrienole (Vitamin E) DGF-Einheitsmethoden F-II 4a, SOP 3.IV.02. ver. 1.).

3 Results and discussion

As a result of tocopherols estimation in lipids of seabuckthorn seeds it was established that α-tocopherol is the major tocopherol isomer. Content of α-tocopherol in Yantarnaya variety was higher (4.8±1.9 mg/100 g) compare to the rest of varieties. Lowest level of α-isomer has been found in Dar Katuni variety (2.1±1.6 mg/100 g). Mean value of α-tocopherol through all varieties is 3.5±0.3 mg/100 g (table 1).

Table 1. Tocopherol isomers content in lipids of seabuckthorn seeds, mg/100 g

Ecotype	Variety	α-toco	β-toco	γ-toco	δ-toco
Buryat	Zarja Dabat	4.2±2.2	0.4±0.3	1.7±1.3	0,3±0.2
Katun	Dar Katuni	2.1±1.6	0.3±0.2	1.0±0.8	0.2±0.1
	Novost Altaya	4.2±3.2	0.8±0.2	1.6±1.2	0.8±0.6
Krasnoyarsk-Sayan	42-68-2	3.8±0.4	0.3±0.0	0.9±0.5	0.3±0.2
	Zhivko	4.4±1.3	0.4±0.2	1.3±1.1	0.2±0.0
Mutants	Elizaveta	2.4±2.1	0.5±0.0	1.1±1.1	1.5±1.2
	Inja	2.2±1.3	0.3±0.1	0.9±0.8	3.5±3.3
Sayan-Katun	Velikan	3.4±1.1	0.5±0.2	1.4±1.2	0.2±0.0
	Yantarnaya	4.8±1.9	0.7±0.4	1.7±1.4	0.1±0.1
Sayan-Chulyshman	Lubimaya	2.6±0.7	0.4±0.2	0.9±0.7	0.8±0.6
	Chulyshmanka	4.2±1.5	0.6±0.2	1.6±1.1	0.3±0.1
Chuya	Chuyskaya	3.3±2.4	0.9±0.5	1.0±0.7	0.6±0.5
$\bar{X} \pm m$		3.5±0.3	0.5±0.1	1.3±0.1	0.7±0.3
min-max		2.1-4.8	0.3-0.9	0.9-1.7	0.1-3.5
V, %		27.2	39.8	26.1	130.6
LSD_{05}		$F_f<F_{05}$	$F_f<F_{05}$	$F_f<F_{05}$	$F_f<F_{05}$

Within years of investigation content of β-tocopherol in seabuckthorn seeds varied from 0.3±0.2 (Dar Katuni) to 0.9±0.5 mg/100 g (Chuiskaya) with mean value at 0.5±0.1 mg/100 g and variation of 39.8%. Content of γ-tocopherol in lipids of seeds ranged from 0.9±0.5

(42-68-2) to 1.7±1.3 mg/100 g (Zarja Dabat) and mean value 1.3±0.1 mg/100 g. Elizaveta and Inja varieties accumulated highest volumes of δ-tocopherol – 1.5±1.2 and 3.5±3.3 mg/100 g respectively. Lowest level of δ-tocopherol was noted in Yantarnaya variety – 0.1±0.1 mg/100 g. Mean amount of δ-isomer is 0.7±0.3 mg/100 g.

As regards comparative tocopherols content estimation between seabuckthorn ecotypes it has been noted that volumes of α-tocopherol ranged from 2.3±1.7 (mutants) to 4.2±2.2 mg/100 g (Buryat ecotype). High level of this isomer has been found in seeds of Sayan-Katun and Krasnoyarsk-Sayan ecotypes. Content of β-tocopherol ranged from 0.3±0.1 (Krasnoyarsk-Sayan ecotype) to 0.9±0.5 mg/100 g (Chuya ecotype). Highest level of γ-tocopherol in seeds has been accumulated by Buryat ecotype, while lowest level by Chuya ecotype – 1.7±1.3 and 1.0±0.7 mg/100 g respectively. Mutants have been distinguished by highest volumes of δ-tocopherol which accumulated up to 2.5±2.3 mg/100 g. There have not been found any significant differences in tocopherol isomers content depending on year of investigation.

Most of results concerning tocopherol estimation that have been found in literature are about total content of tocopherols but not isomers one. That is why it is quite difficult to compare obtained results with represented in literature.

For instance according to [4] tocopherols content in seeds of Altai seabuckthorn is 14.3 mg/100 g. At the same time according to most recent investigations [5] tocopherols content in seeds of seabuckthorn is much higher and reported as about 62.15 mg/100 g.

Some foreign researches like Beveridge et al. [6], Zhang et al. [7] reported about total content of tocopherols in seed oil from 61.0 to 113.0 mg/100 g. According to Kallio et al. [8] tocopherols content is about 28.7 mg/100 g, that much lower compare to previous authors. In 2017 group of researches from Slovakia carried out more comprehensive investigation of tocopherols content in seabuckthorn seeds. It has been reported that major tocopherol is α-tocopherol, that in accordance with our results. The amount of α-tocopherol in seabuckthorn seeds is about 70.0 mg/100 g. β-, γ- and δ-isomers accumulated in lower amounts. For instance β-tocopherol content was 17.07 mg/100 g, γ-tocopherol – 45.85 mg/100 g, δ-tocopherol – 9.47 mg/100 g [9]. These results are much higher compare to that obtained by our research. Because of lack of information on this topic, quite different results of them the further investigations and analyses are necessary.

References

1. N.M. Emmanuel, Y.N. Lyaskovskaya, *Inhibition of lipid oxidation processes* (Permagon Press, New York, 1961)
2. A.V. Trufanov, *Biochemistry and physiology of vitamins and antivitamins* (Moscow, 1959)
3. A. Abbasi, *Functional analysis of tocopherol biosynthesis in plants* (Aus Tehran: IRI, 2007)
4. D.A. Obodovskaya, *A seabuckthorn as raw material for the vitamin industry* (Moscow, 1957)
5. A.M. Zolotareva, M.G. Zayatueva, N. Bolorcecag, S.B. Rinchinov, J. Chemistry of plant raw materials **3**, 177-183 (2015)
6. T. Beveridge, T.S.C. Li, B.D. Oomah, J. Agric. Food Chem. **47**, 3480-3488 (1999)
7. W. Zhang, J. Yang, J. Duo, B. Ren, J. Guo, *Proceedings of International Symposium on Sea Buckthorn* (*H. rhamnoides* L.) 96-105 (Sian, 1989)
8. H. Kallio, B. Yang, P. Peippo, R. Tahvonen, R. Pan, J. Agric. Food Chem. **50**, 3004-3009 (2002)

9. Z. Burcovaa, F. Krepsa, S. Schmidta, M. Jablonsky, A. Haz, A. Sladkova, I. Surina, J. Acta Chimica Slovaca **10**, 29-34 (2017)

BIO Web of Conferences **11**, 00052 (2018) https://doi.org/10.1051/bioconf/20181100052
Prospects of Development and Challenges of Modern Botany

Comparative analysis of trypsin inhibitor activity in the wet and dry weight of leaves in representatives of *Hedysarum* L. in the forest-steppe of Western Siberia

Elena Zhmud[*], *Irina* Kuban, *Maria* Emtseva, and *Olga* Dorogina

Central Siberian Botanical Garden Siberian Branch Russian Academy of Science, 630090, Zolotodolinskaya str., Novosibirsk, Russia

Abstract. The trypsin inhibitor activity (TIA) in fresh leaves in *Hedysarum alpinum* L., *H. flavescens* Regel & Schmalh. and *H. theinum* Krasnob. in Central Siberian Botanical Garden Siberian Branch Russian Academy Science (CSBG SB RAS, in the forest-steppe of Western Siberia) was investigated in the vegetative phase of seasonal development. Trypsin inhibitors (PR-6 proteins) play the role of immunity factors, since under various influences on plants, the genes encoding the production of this group of substances are activated. They are synthesized in different plant organs, and are necessary for plant control with biotic and abiotic stresses. TIA in the species was characterized by low (< 20 mg/g dry weight) and medium (≥ 20 mg/g dry weight) values, as we identified. It was found that the TIA values vary at a high and very high degree when determining in fresh leaves, and in the study in dry leaf flour - to medium degree. The preservation of TIA values in the species in the dried leaves is shown. On the basis of this fact, the possibility of studying TIA in these species in dried leaves is justified.

1 Introduction

It is known, that various parts of all angiosperms contain substances that promote the protection of the plants from damage and phytophages. One of the groups of such substances are inhibitors of proteinases, and in particular, trypsin inhibitors (TI). For practical purposes in storage organs in many cultivated plants these substances are isolated and their properties are studied. These are immunity factors, since different influences on plants causes activation to occur on the genes encoding the production of this group of substances, performing protective functions. Enzyme systems are activated, membrane

[*] Corresponding author: : elenazhmu@ngs.ru

stabilization occurs, the activity of mitochondria and chloroplast functioning increases, and accordingly energy supply due to stress proteins [1]. The loss of crops due to various biotic and abiotic factors is one of the urgent problems of modern agriculture. A multicomponent protection system: the creation of protective barriers, the activation of a hypersensitivity reaction, the synthesis of antimicrobial low-molecular peptides that exhibit a wide range of activity against fungi, bacteria and viruses, such as PR-6 proteins in particular, or trypsin inhibitors (TI), exist in the plants. Modern methodological approaches allow effective search for new forms of activity of such proteins in wild non-model plant objects [2]. Protease inhibitors showed antitumor activity, capable of blocking the ability of tumor cells in the formation of metastases. For the treatment of Acquired Immunodeficiency Syndrome (AIDS) and hypertension, some protease inhibitors are also used. Other studies have shown the efficacy of these molecules as an antiprotozoal drugs and for the treatment of diabetes and Alzheimer's disease [3,4].

Hedysarum alpinum L., *H. flavescens* Regel & Schmalh. and *H. theinum* Krasnob., are polycarpic caudexo-forming monocentric herbaceous plants with elongated monocarpic shoots. For the treatment of cardiac, pulmonary, catarrhal and skin diseases of the viral nature, these species are used. Their medicinal properties are due to a unique set of biologically active substances [5]. In addition, in the leaves of these plant species, we found high values of trypsinin inhibitory activity (TIA), which can serve as a basis for the isolation and practical use of this group of biologically active substances [6-8].

TIA is studied mainly in seeds in cultivated plants, according to literature data. Discussion of TIA conservation in fresh and dry plants leaves Fabaceae is not found in literature. At the same time, the determination of the activity of this group of substances in plants under natural conditions, remote from the location of laboratory research, is important in solving a number of problems.

The purpose of the article is to show the results of a comparative study of TIA in leaves (fresh and dried) in species of *Hedysarum alpinum, H. flavescens, H. theinum* in culture in the forest-steppe of Western Siberia.

2 Materials and methods

TIA in *Hedysarum alpinum, H. flavescens* and *H. theinum* was studied in Central Siberian Botanical Garden Siberian Branch Russian Academy Science (CSBG SB RAS), in the forest-steppe of Western Siberia. Mature one-aged (8 years) generative individuals were studied. Values are expressed in milligrams TIA pure bovine trypsin inhibitor bound per 1 g of air-dried leaves flour (mg/g dry weight, more mg/g). BAPA (Na-benzoyl-DL-arginine-p-nitroanilide) was used as a substrate for the study [9], as well as bovine trypsin (reagents manufactured by ISN-Biomedical, USA)). The method is adapted by us to study the TIA in the leaves of plants on the basis of methods for determining the TIA in the leguminous seeds [10]. This method is based on the spectrophotometric measurement of the optical density of the decomposition products of the protein substrate (BAPA) under the action of trypsin (a wavelength of 405 nm). Adding plant extract with inhibitors, binding trypsin, was accompanied by a decrease of extinction. As the buffer, 0.05 M Tris–HCL–0.02 M $CaCl_2$ (pH=7.7) was used. TIA was considered low at values when less than 20 mg/g, average - from 20 to 40 mg/g. TIA in fresh leaves was studied simultaneously in 10 individuals of each species in the autumn vegetative phase (9.08.2017). The compound leaf of each individual of the middle part of the generative shoot was placed in a humid chamber and transferred to the laboratory conditions. Then it was weighed before and after taking a leaf for analysis. The fresh leaf was homogenized in 5 ml of distilled water and left overnight at + 40°C for extraction, after which the TIA was determined. The remainder of the sheet was dried in laboratory conditions for 30 days (6% moisture) (hereinafter - dry substance), then made into flour and sieved through a sieve of 0.1 mm cell. Extraction IT was also performed in distilled water overnight at + 4^0C. Statistical processing was carried out using the StatSoft STATISTICA software package. Reliability was taken into account using the non-parametric U-test criterion Wilcoxon-Mann-Whitney [11]. Variation was evaluated on the scale S.A. Mamaev [12].

3 Results and discussion

In the vegetative phase of development in the fresh leaves of the plants, TIA reached low and middle values, which are not statistically significant differences, as shown by our study. Recalculation TIA values on a dry weight basis was conducted after the study in the fresh leaves. The mean degree of leaf shrinkage values for *Hedysarum alpinum, H. flavescens* and *H. theinum* had comparable values and amounted to 2.9 ± 0.1, 3.1 ± 0.1 and 3.2 ± 0.2, respectively. TIA in *H. alpinum* and *H. theinum* samples were characterized by low values, and the mean values of this trait in *H. flavescens* were noted (Table 1).

Table 1. TIA in leaves of some Hedysarum species; 1 - in fresh leaves;
2 - in fresh leaves, recalculation to dry matter; 3 - dry matter determination.

Indicators	*H. alpinum*			*H. flavescens*			*H. theinum*		
	1	2	3	1	2		1	2	3
M	6.7	19.2	14.7	5.8	23.7	24.5	4.5	13.3	15.0
m	1.7	4.4	1.2	1.9	7.9	4.1	1.0	2.4	1.4
V,%	57.7	50.8	17.5	72.1	75.1	37.5	50.1	40.0	20.8
min	4.1	12.4	11.5	1.7	5.8	9.8	1.7	6.7	11.5
max	13.4	35.8	18.0	10.6	46.6	34.4	7.2	19.9	19.6

Note. M - Mean value; m - Error; V - Coefficient of variation, %; min, max - Range of values.

The variability of calculated TIA values for the sample in terms of dry matter was higher than in the direct measurement of the characteristic in dry matter. Species with low values of about 2.9 and 1.7 times, in a species with mean values of sign 8.3 and 3.5 times it was respectively. Thus, we have shown that the values of the studied samples TIA vary differently, that is, depending on the initial state of the investigated samples. The variation coefficient was high and very high when studying the trait in fresh leaves in all samples. This coefficient varied within average values when studying TIA in dry leaves (Table 1). The study of TIA in fresh material is not possible when working in natural conditions with wild plants. Our study showed that TIA statistically significant ($p < 0.05$) is retained in the dry matter of leaves in the studied species *Hedysarum alpinum, H. flavescens* and *H. theinum* (Fig. 1).

Saving the antioxidant activity of oxidases and peroxidases was obtained, for example, *G. procumbens* air dried at ambient temperature, whereas the drying at 50°C and 100°C substantially reduced the antioxidant enzymes, according to the literature [13]. Using the freeze drying method, a significant decrease in the activity of all the enzymes studied was revealed [14]. With the use of gentle mechanical drying parameters, the preservation of the proteolytic activity of the pineapple extract was detected [15]. Comparative analysis in the studied species of Hedysarum showed conservation of TIA in dry matter. TIA is more suitable for studying in dry matter, since the activity of this group of substances is statistically significant keep in in wet and in terms of dry weight, and the results for this analysis are much less variable than meanings the TIA in fresh leaves.

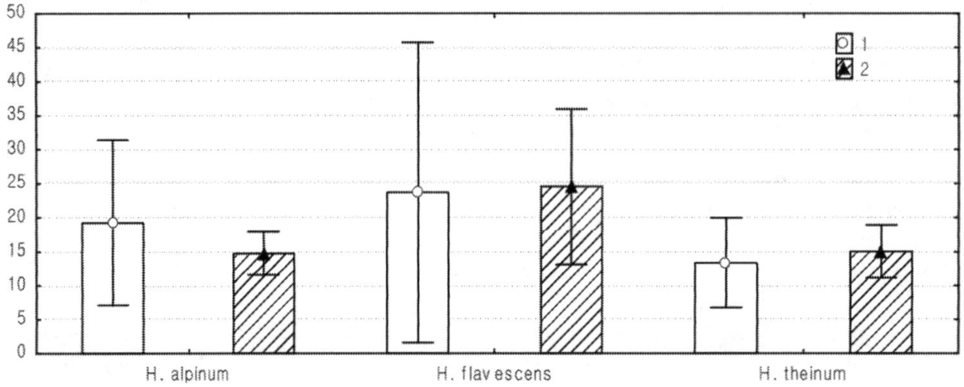

Fig. 1. TIA in raw and dry matter in the species of the genus Hedysarum L. (CSBG SB RAS, 9.08.2017). Designations: 1 - TIA in fresh leaves in terms of dry matter (mg/g dry weight); 2 - TIA in dry leaves flour (mg/g dry weight). On the X-axis - Mean values of the characteristics with 95% confidence interval; on the Y-axis - species names, studied.

References

1. N.N. Karpun, E.B. Yanushevskaya, E.V. Mihajlova, Agricult. Biol. **50**, 540–549 (2015)
2. O.A. Kulaeva, M.S. Klyukova, M.N. Povydysh, I.A. Tihonovich, V.A. Zhukov, Agricult. Biol. **52**, 3–12 (2017)
3. P.M.G. Paiva, E.V. Pontual, L.C. Coelho and T.H. Napoleão, Formatex 641–649 (2013)
4. G. Fear, S. Komarnytsky, I. Raskin, Pharmacol. & Therapeut. **113**, 354–368 (2007)
5. N.S. Zinner, T.A. Kukushkina, G.I. Vysochina, T.P. Sviridova, Vestn. Tomsk. Gosud, Univ., Biol. **4,** 116–122 (2010)
6. O.V. Dorogina, E.V. Zhmud, Sibirs. Vest. Sel'sk. Nauki **10**, 23–28 (2010)
7. E.V. Zhmud, N.S. Zinner, Rastit. Resursy **3**, 103–111 (2011)
8. E.V. Zhmud, N.S. Zinner, O.V. Dorogina, Vest. Tomsk. Gos. Univ., Biol. **3**, 100–110 (2012)
9. Y.Y. Gofman, I.M. Vajsblaj, Priklad. Biohim. i Mikrobiol., **11**, 777–783 (1975)
10. A.I. Ermakov (ed.) *Metody biohimicheskogo issledovaniya rastenij* (Kolos, Leningrad, 1987)
11. Wilcoxon–Mann–Whitney U-test http://medstatistic.ru/theory/mann.html
12. S.A. Mamaev, *Ekologiya* **1**, 39–49 (1975) (in Russian)
13. J.J. Fung, Y.Y. Lim, Reinvention: An Int. Journ. of Undergrad. Resear. **9**, 2 (2016)
14. B. Moura, M.S. Neidiquele, E.C. Machado, R.V. Ribeiro, Brazilian Journ. of Bot. **39**, 1 (2015)
15. A.C.S. Cabral, S. Said, W.P. Oliveira, Int. Journ. of Food Proper. **12**, 536–548 (2009)

AUTHOR INDEX

AUTHOR INDEX